DATE DUE

AP 24 '00		
DE 19 '00		
AP 18 01		
NO 24 09		

ACCIDENT PREVENTION

and

OSHA COMPLIANCE

Patrick A. Michaud
Safety Consulting Service
Sanford, Maine

LEWIS PUBLISHERS

Boca Raton   New York   London   Tokyo

Library of Congress Cataloging-in-Publication Data

Michaud, Patrick A.
    Accident prevention and OSHA compliance / by Patrick A. Michaud.
      p.   cm.
    Includes index.
    ISBN 1-56670-150-3
    1. Industrial safety--United States.   2. Industrial safety--Law
and legislation--United States.   I. Title
    T55.M486   1995
    363.11′56′0973--dc20   94-44946
                                                                                   CIP

© 1995 by CRC Press, Inc.
Lewis Publishers is an imprint of CRC Press

No claim to original U.S. Government works
International Standard Book Number 1-56670-150-3
Library of Congress Card Number 94-44946
Printed in the United States of America 1 2 3 4 5 6 7 8 9 0
Printed on acid-free paper

PREFACE

For many years I have searched for the single safety book that would give busy managers an understanding of accident prevention — a book that would tell them how to reduce their losses, how to comply with the many OSHA regulations, and how to avoid an OSHA inspection.

Even after the Occupational Safety and Health Act of 1970, when thousands of new regulations were written to help the managers, the safety and health requirements were vague and subject to interpretation by industry and OSHA itself.

Safety professionals wrote many books on their version of safety and health. This further confused owners and managers of businesses to the point that accident prevention was relegated to a safety manager who handled the job and told the boss only what he wanted to hear. When the going got bad, the boss blamed the safety manager and often looked for a new person to take over.

As a safety consultant for industry, I have been asked by managers, "Why can't someone write an accident prevention book that I can understand?" After hearing this a number of times, I decided to write a book for those managers. My book explains accident prevention and how to avoid an OSHA inspection. Why? Because most managers want to reduce their losses — especially when they are preventable. They are also looking for new ways to compete in the marketplace at a time when they face new and more aggressive world markets.

Accident prevention can be just as effective as any other production controls such as quality, theft, and drugs. No one will deny that those controls are necessary in order to produce a product that will compete with others.

The same applies to accident prevention, because without some control, you are bound to suffer more losses. The easy-to-apply principles make this a ready reference book on how to prevent accidents, injuries, illnesses, fires and other disasters. The chapters are written in everyday English without buzz words. The methods presented are sensible, businesslike, and they work.

Yes, you can avoid an OSHA inspection if you believe that accidents are preventable. I thank all the managers that I have come to know, and I give much credit to the Department of Labor (OSHA) for the use of their regulations and their booklet publications.

Patrick A. Michaud
Sanford, Maine

THE AUTHOR

 **Patrick A. Michaud** is the owner of Safety Consulting Service of Sanford, Maine. He was formerly the Safety and Health Manager of Portsmouth Naval Shipyard, Portsmouth, New Hampshire. Mr. Michaud began his career in the construction field as a supervisor. He then went on to shipbuilding, holding various positions in mangagement, design and safety. He is a registered professional engineer, a certified hazard control manager at the master's level, and is currently a safety consultant. Mr. Michaud has taught safety and health for many years at the University of Southern Maine.

Mr. Michaud is listed in *Who's Who in America* as a safety professional. He is active with safety groups, the Maine Highway Safety Commission, the Maine Prevention of Blindness, local fire and rescue organizations, and is a former board member of York County Health Services, and a member of the American Industrial Hygiene Association.

Mr. Michaud has written over 100 safetygrams, numerous articles for the *Navy Lifeline Magazine* and a considerable number of safety/health instructions for the Navy. He has been involved in safety since 1950, when he served in a damage control unit for the United States Navy.

He has a Bachelor of Science degree from the University of Maine and is a graduate of Southern Maine Vocational and Technical Institute in the Technology of Construction.

HOW THIS BOOK WILL HELP YOU TO IMPROVE YOUR SAFETY AND HEALTH PROGRAM AND REDUCE YOUR LOSSES

Whether you own a large corporation or a small business, your greatest asset is your employees. This book will help you to understand why you are losing your profits due to accidents, which can result in equipment loss and employee injury. You will find out how you can avoid or survive an OSHA inspection not by doing more, but rather by doing things right the first time. This book will tell you where your problems are, and in easy-to-understand language, describe the steps you need to take to make OSHA happy. When you have done this, you will be even happier than OSHA, because you will have found the secret to plugging the profit leak.

You will find this book to be practical and handy with short, easy-to-apply chapters designed to match your problems to the proper chapter that will resolve them. You will find that it:

- defines the responsibilities for safety (Chapter 1)
- explains how to maintain a safety program (Chapter 1)
- tells why safety needs no apology (Chapter 1)
- explains the philosophy of safety (Chapter 2)
- shows why people resist change (Chapter 2)
- lists the ten points of a safety policy (Chapter 2)
- defines the three accident causes (Chapter 3)
- tells about unsafe acts and conditions (Chapter 3)
- shows how to correct unsafe acts and conditions (Chapter 3)
- explains how to look for hazards (Chapter 4)
- discusses how to check each step of a job (Chapter 4)
- shows how to construct a safety inspection list (Chapter 4)
- discusses the definition of an accident (Chapter 5)
- tells how to investigate accidents (Chapter 5)
- explains the benefits of accident investigations (Chapter 5)
- explains how to conduct a job hazard analysis (Chapter 6)
- identifies each step of the job and the hazards (Chapter 6)
- explains how to revise a job hazard analysis (Chapter 6)
- tells about the importance of safety training (Chapter 7)
- explains how to train your employees (Chapter 7)
- discusses on-the-job training (Chapter 7)
- explains the study of human factors (Chapter 8)
- tells about human limitations (Chapter 8)
- shows how to use the handicapped workers (Chapter 8)

- tells about fire safe construction (Chapter 9)
- describes a fire safety check list (Chapter 9)
- lists the different types of fire causes (Chapter 9)
- explains the exposure of health hazards (Chapter 10)
- discusses the professional health evaluation (Chapter 10)
- identifies the human body sensors (Chapter 10)
- explains the Hazard Communication Law (Chapter 11)
- shows how to train the employees in hazard communication (Chapter 11)
- tells how to protect a trade secret (Chapter 11)
- discusses the many types of respirators (Chapter 12)
- lists the different medical restrictions (Chapter 12)
- explains local exhaust ventilation (Chapter 12)
- describes personal protective equipment (Chapter 13)
- explains why protection should be worn (Chapter 13)
- identifies the parts of the body to be protected (Chapter 13)
- discusses the total safety program (Chapter 14)
- tells how to anticipate accidents (Chapter 14)
- explains how to learn from mistakes (Chapter 14)
- explains the purpose of emergency response (Chapter 15)
- identifies the emergency action team (Chapter 15)
- tells about the importance of training for emergencies (Chapter 15)
- explains the purpose of safety committees (Chapter 16)
- shows how to organize a safety committee (Chapter 16)
- tells how to conduct a safety committee meeting (Chapter 16)
- discusses the requirements of record-keeping (Chapter 17)
- tells how to report accidents to OSHA (Chapter 17)
- explains how to compute an incidence rate (Chapter 17)
- illustrates the direct and indirect costs of accidents (Chapter 18)
- provides examples of direct and indirect costs (Chapter 18)
- shows how to compute the cost of an accident (Chapter 18)
- tells how to reward the employee for safe work practices (Chapter 19)
- explains how safety and productivity go hand-in-hand (Chapter 19)
- lists the various types of safety awards (Chapter 19)
- shows how to enforce safety rules (Chapter 20)
- explains disciplinary action and alternate solutions (Chapter 20)
- discusses supervisory control over employees (Chapter 20)
- explains first aid and medical care (Chapter 21)
- shows how to set up a first aid station (Chapter 21)
- tells about required physical examinations (Chapter 21)
- explains how the law protects employees (Chapter 22)
- discusses the OSHA inspection process (Chapter 22)
- lists the different OSHA citations and fines (Chapter 22)
- explains the effects of drugs and alcohol (Chapter 23)
- shows how to recognize the "user" (Chapter 23)
- identifies the user as an accident maker (Chapter 23)
- describes various office hazards (Chapter 24)
- lists the types of office injuries (Chapter 24)
- tells how to maintain a safe office (Chapter 24)

- lists the common home and play injuries (Chapter 25)
- explains how to make your home safe (Chapter 25)
- discusses the benefits of "safety always" (Chapter 25)
- lists the common preventable injuries (Chapter 26)
- explains how to protect the employees (Chapter 26)
- describes the cost of preventable injuries (Chapter 26)
- tells how to avoid a tragic work fatality (Chapter 27)
- describes the six deadly accidents (Chapter 27)
- shows how to prevent work-related fatalities (Chapter 27)
- explains the purpose of self-evaluation (Chapter 28)
- shows how to set goals and objectives (Chapter 28)
- tells how to build safety into every operation (Chapter 28)

Some will tell you that it is impossible to avoid an OSHA inspection because an employee can report you to OSHA by alleging that an unsafe condition exists in your workplace. Yes, that can happen, but remember, OSHA will only investigate that complaint, and if the complaint is not valid, OSHA will not stay to inspect your workplace. Also, when you have subscribed to every chapter of this book, your employees will have no reasons to report you to OSHA.

CONTRIBUTORS

Occupational Safety and Health Admininstration
Various Regulations and Publications

National Institute for Occupational Safety and Health
Publication 78-197

American National Standards Institute Inc.
Eye and Face Selection Guide

Penetone Corporation
Charts, Tables, and Formulas

American Society of Safety Engineers
Terms Used in the Safety Profession

International Labour Organization
Short Excerpts

American Red Cross Association
Information Regarding Drugs for Chapter 23

National Fire Protection Association

National Safety Council

Note: Where permission was required to use material, it was granted in writing.

IN APPRECIATION AND GRATITUDE

A number of agencies provided material for this book. Therefore, my heartfelt thanks to them and to those who gave me their support.

- Mary Ann Michaud, my mother and a school teacher
- Ginny Maggio for her most patient manuscript typing with numerous revisions
- Ronald Dugre and Richard Michaud
- The American Red Cross
- Bureau of Labor Statistics
- Occupational Safety and Health Administration (OSHA)
- National Institute for Occupational Safety and Health (NIOSH)
- American Safety Council
- American Society of Safety Engineers
- Penetone Corporation
- National Fire Protection Association
- American National Standards Institute
- National Institute on Drug Abuse

CONTENTS

DEDICATION

*To Pauline, my wife and best friend, who helped me organize
the contents of my manuscript.*

1   INTRODUCTION

"A man always has two reasons for doing anything — a good reason and the real reason."

J.P. Morgan

AN OUTSTANDING ACHIEVEMENT

Years ago, when man stepped on the surface of the moon for the very first time, the world watched that achievement with great pride and much interest. That accomplishment was not due to chance, but rather to the work and planning of the professionals who were engineers and scientists. The event was indeed spectacular in the scientific sense, but also in the moral sense (International Labour Office, Geneva, Switzerland). The success of the mission could not have happened without a program to ensure the safety of the astronauts and the safety systems of the space vehicle.

IMPORTANT CONCERNS

Today, concerns are being expressed about pollution and damage to our environment. We are nervous about the thousands of hazardous chemicals and compounds that are produced with little thought to their handling and disposal. Admittedly, we only really got excited about physical injuries when the Occupational Safety and Health Act was signed into law in 1970. That law told employers to provide a healthful workplace for their employees; the assurance that they would comply with the Act would be the Bureau of Labor's statistics that flagged their injuries and the OSHA inspections that would follow.

OUTSTANDING SAFETY PROGRAMS

Successful companies throughout the world all have outstanding safety and health programs, which again, did not happen by chance. They recognized that the success of their mission would be enhanced by the professional management of safety. Today's managers will have to provide aggressive and consistent leadership. They will have to be innovative in the application of standards and controls. And, they will have to provide adequate financial and manpower resources to control the hazards that interrupt the order of good business. Without management's commitment, progress is unlikely and injuries and illnesses in the workplace will continue.

SAFETY FIRST!

Safety is not and cannot be first. Instead, what is first is profit, mission, product, service, and the strength and growth of your business. However, safety is important because it can either support or damage the above, so it must be fully integrated into your plans and operations. Safe and efficient production will maximize your profits. In other words, safety is the avoidance of loss.

Safety is no different than your other loss control programs. No manager involved in the production of a product would ignore certain controls such as quality, fire, security, absenteeism, drug and alcohol intake, etc., yet in some companies, safety receives only token recognition. When this happens, those companies are often driven out of business.

RESPONSIBILITY FOR SAFETY

It has been said by business managers, "that accidents and their associated problems are a failure of management."

Abraham Lincoln once said, "Next to creating life, the greatest thing man can do is to save a life." He also said, "It is the duty of every man to protect himself and those associated with him from accidents, which may result in injury or death."

When I am asked to give a good reason why managers should insist on a meaningful safety program, I answer with certainty:

- to reduce injuries
- to reduce human suffering
- to reduce property damage
- to reduce compensation claims
- to increase production and profits
- to increase morale
- to improve job satisfaction
- to improve housekeeping

The Occupational Safety and Health Act of 1970 tells all employers that they are indeed responsible for the well-being of their employees. There is no question that we are all responsible for safety, and it is for this reason that I will mention an article that I found years ago by an anonymous writer.

This is a story about four people named Everybody, Somebody, Anybody, and Nobody. There was an important job to be done and Everybody was asked to do it. Everybody was sure Somebody would do it. Anybody could have done it, but Nobody did it. Somebody got angry about that, because it was Everybody's job. Everybody thought Anybody could do it, but Nobody realized that Everybody would not do it. It ended up that Everybody blamed Somebody when Nobody did what Anybody could have done.

The story tells it all. Safety is Everybody's job, from the top person of your company to the very bottom person. Unless and until top management in any business recognize and understand their obligations in the area of occupational safety and health, there will be little, if any, progress.

When you look at the cost of accidents in terms of death, disability, illness, pain, lawsuits, delays and the loss of a worker, you will have to agree that there is only one answer — safety always.

MAINTAINING SAFETY

You can maintain safety consciousness by giving your employees safety lectures, insisting that they memorize safety rules, by putting up safety posters, and rigidly enforcing safety rules. These measures will have some effect but they will not do the whole job; they need to be reinforced by management. Why? Because you have to protect people from themselves. That is the key to maintaining a safety program.

For example, we all tend to lose our fear of the tools and equipment we work with, and we forget the hazard of the job we are working on. Familiarity soon breeds contempt for the dangers that surrounds us — and we often unconsciously create unsafe conditions and situations that lead to accidents while thinking "it can't happen to me".

People need to be jolted into realization that in a single second it can happen — an accident! What is the best way to drive this point home? Make an effort to change your workers' attitudes by first changing your supervisors' attitudes. Remember this, "The world's best safety device is a careful worker and a concerned supervisor."

SIZE OF SAFETY PROGRAMS

Safety programs and policies must be tailored and fitted to your objectives. There is no absolute model to follow as every company is different in size and mission, but the methods used can all be successful if they contain the elements as outlined in the chapters of this book. Within smaller companies, it may be more difficult to follow the lead of a larger company that employs many full-time safety and health professionals. However, as a small business, you can get help from a consultant and/or you can assign the role of safety manager to your plant engineer on a collateral duty basis.

When you do this, that person should report directly to you. There are many companies that operate successful safety programs this way. Whoever you assign to carry out your safety program, that person should be given time to do the job. It is a good idea to expose that person to some safety training and the reasons why safety is good business.

For larger companies, you should have a well-trained safety group. The group should be led by a qualified professional who is able to direct a safety and health staff. The staff's primary function would be to assist your supervisory force in maintaining safe and healthful working conditions. They would also instruct your employees in general safety rules and procedures.

SAFETY NEEDS NO APOLOGY

During my 40 years in the construction and the industrial business, I have heard some managers and supervisors apologize for safety by saying "OSHA tells us to do it this way. I don't agree, but we have to do it", or in giving a safety talk, a foreman addressed his workers by saying "I have to talk to you about safety. This will only take a minute." This is like saying we are wasting time, but we have got to do it. That type of attitude will have a domino effect throughout the entire structure of your business. The attitudes of your employees will be ones which pull your safety program apart.

No one should ever apologize for safety. Safety is the professional way to help you to accomplish your mission without the loss of a life, injury to your people, damage to equipment, and the loss of your business.

SUMMARY

The world's most successful companies all have outstanding safety programs. Their safety programs did not just happen! With these companies, safety is a condition of employment. In your own company, supervisors should be charged to make certain that their workers observe safety rules. Safety should never be considered a part-time effort, but instead a full-time job for everyone from the top person of the company right down to the bottom person. Remember this, accidents are usually caused by people, and seldom by properly maintained tools, equipment, or machinery.

2   ACCIDENT PREVENTION

"Action is the proper fruit of knowledge."

Thomas Fuller

SAFETY PHILOSOPHY

Some time ago, the owner of a small industrial firm mentioned to me that his company was experiencing a bad trend involving accidents and injuries. He wanted to know what could be done to reverse that unacceptable trend. After looking over his safety program at the paper level, then on down to the production area, I told him that the first step should be to establish a management philosophy toward safety and that he must believe that all accidents are preventable. Second, all accidents and injuries must be investigated in order to prevent their recurrence. Third, that all unsafe conditions and unsafe acts must be corrected immediately.

I also told him that good workers represent an investment in both time and money. Their absence from the labor force for any amount of time and for any reason results in production loss. When that absence is caused by an accident that could have been avoided through the elimination of an unsafe condition, the wearing of personal protective equipment or safe work habits, the loss is doubly unfortunate.

I stressed that supervisors are the key to a good accident-free program. They are expected to lead workers and not allow them to take chances that could cause bodily harm. Workers, on the other hand, are expected to follow instructions — doing the job safely and on time. No one should be required to compromise safety for the sake of more production, nor production for the sake of safety. In other words, a good job is a safe job and a safe job is a good job.

RESISTANCE TO CHANGE

People usually have mixed feelings at about a relationship or condition. They want to be safe, but they also enjoy flirting with danger. They want to be progressive, but they resist change. Employees' relations to change are very perplexing and this applies to safety rules as well as to any other condition of employment. Because of this, your supervisors must recognize that this common, but at times ignored trait of human nature; the resistance to change — that resistance has a significant influence on your workers' behavior and attitudes toward safety.

There is much that you can do to motivate your employees to accept change. Tell your employees in advance that you plan to enforce safety and how that will affect them. Your supervisors should explain the change because people resist what they do not understand. However, it is not enough to explain *what* the change is, your employees must also be given the *reasons* for the change, (1) to prevent injury, (2) to prevent loss of work, and (3) to avoid hospitalization. Also, it is what your workers see in a situation that governs their actions and reactions.

Remember, management does set the style. When management shows through their safety practices that they really believe in good safety concepts and methods, that attitude will be reflected in your employees' safety behavior as well. The reverse is also true, when management gives nothing more than lip service to safety, fails to use safety equipment that they themselves expect their workers to use, or tolerates poor housekeeping and unsafe work practices, your employees will have an "I could care less" attitude about safety, and at that point, your accident rate is bound to increase.

SAFETY POLICY

As the owner or manager of a business, you must have a written policy regarding safety. An established safety policy signed by yourself and your top managers and made known to your employees does more than assure your employees that your management team is interested and committed to their safety. It also shows that you are willing to take all practicable means to make your workplace safe. A safety policy also assures your supervisors that their efforts to promote safety are supported by top management and are recognized as a necessary function in efficient production.

The following written safety policy, firmly stated and conscientiously followed, will go far to enlist the cooperation of your workers and supervisors.

1. The most valuable resource is the worker, therefore, the worker must be protected from harm.
2. All injuries and illnesses can be prevented.
3. Working safely is a condition of employment.
4. Management must audit the safe performance of workers.
5. Management must provide safety training for each task to be performed.
6. Unsafe conditions and unsafe acts must be corrected promptly.
7. Disciplinary action must be constructive and consistent.

8. All accidents, injuries, illnesses, and near-misses must be investigated promptly.
9. Employees must be encouraged to discuss the hazards associated with their work.
10. Management will set the right example by observing all safety rules.

MUCH MORE ABOUT SAFETY

Unless you read and subscribe to every chapter in this book, nothing will change. As you read you will see that each chapter depends on the one before it and the one that follows. Each chapter topic is an important element of a safety program. It is *not* my intention that each chapter serves as the final word regarding each and every subject, but rather that it gives you enough information to stimulate you to read other works.

Accidents do not happen by themselves, they are caused. Unless your people know how to determine the cause of an accident, they will not be able to conduct accident investigations. Nor will they be able to perform safety inspections when they do not know what to look for. Without safety training for your employees, the same types of accidents will occur again and again.

Your safety training program will be ineffective unless your instructors complete a job hazard analysis for all the hazardous tasks. The study of human factors for repeat motions is directly related to job hazard analysis. Fires can be prevented by safety inspections and constant watchfulness.

Occupational health, hazard communication and respiratory protection all depend on each other to protect the workers' health. Consequently, the chapter on personal protective equipment applies to all the previous chapters.

System safety is achieved when all the procedures in all of the preceding chapters are applied in the spirit of safety. The chapter on system safety details the conscientious effort to control the hazards associated with everything you do from start to completion of a project or task. It is a thought process and a check list to assure you that costly mistakes and accidents will be controlled or eliminated.

The chapter on emergency response provides a fallback position so that if something should go wrong, you will be able to respond to any emergency in order to minimize employee injury and property damage.

The chapter on safety committees will also enhance your safety program. While record-keeping and statistics will provide a barometer of your safety efforts, accident cost will tell you how much you could save.

The chapter regarding incentive programs should help foster good safety. Enforcing rules is a must for any safety program. The chapter on first aid and medical personnel will show you how to treat injuries.

The OSHA chapter will tell you in a nutshell what to expect from an OSHA inspection. Hopefully, this will not happen, especially if you read this book and subscribe to its contents.

The chapter on drugs and alcohol will explain why it is so important to recognize and treat the user.

It is certain that the chapter regarding office safety will enhance your safety program by reducing your total injury statistics.

The chapter on off-the-job safety will support your safety program on a 24-hour basis.

The two chapters titled "Preventable Injuries" and "Tragic Accidents" should convince you that those types of accidents are very demoralizing to your workers. The last chapter will provide you with ideas for performing an evaluation of your injury and illness prevention program.

The useful contacts, useful information, tables, glossary, and OSHA information in the appendices of this book will provide you with sufficient knowledge to understand what needs to be done to put your program on the right track.

SUMMARY

A safety policy is a must for any successful safety program. It is not only necessary that you institute a safety policy for your company, but you must also make certain that the safety rules are enforced by your supervisor. It will not mean anything if your employees see that in practice your supervisors do not enforce your policy or safety rules. Or worse, when the supervisor looks the other way.

Your supervisors are the key people that can make or break your safety program. You must insist that all your people, from those at the top on down in your organization comply with your safety policy and all other subsequent safe work practices.

3   ACCIDENT CAUSES

"Everything has a cause and the cause of anything is everything."

W.J. Turner

THE THREE ACCIDENT CAUSES

The cause of anything could be the result of an unsafe condition, or an unsafe act, or an act of God. Accidents are usually caused by one of the above or by all three, but rarely by acts of God. It is a known fact that the majority of all accidents are caused by the unsafe acts of people.

Unsafe conditions are also known as physical hazards and they exist in poorly organized or improperly planned situations; usually because of error or lack of knowledge on the part of those responsible. A good example is a handrail installed at the wrong height, which in turn, can cause people to lose their grip when using stairways. Another physical hazard is when a workbench is installed at an improper height, which can result in back pains.

Unsafe acts are also known as human hazards, and they exist in the minds of supervisors and employees for many reasons. Some examples of human hazards are the result of poor instructions, lack of ability, poor discipline, poor attitudes, emotional unfitness, or just plain taking shortcuts.

UNSAFE CONDITIONS

The definition of an unsafe condition, according to The American Society of Safety Engineers is

> ...any physical state which deviates from that which is acceptable, normal, or correct in terms of its past production or potential future production of personal injury and/or damage to property or things; any physical state which results in a reduction in the degree of safety normally present. It should be noted that accidents are invariably preceded by unsafe acts and/or unsafe conditions. Thus, unsafe acts and/or unsafe conditions are essential to the existence or occurrence of an accident.

Unsafe conditions are

- Inadequate guards
- Congested work-sites
- Inadequate warning devices
- Fire and explosion hazards
- Poor housekeeping
- Hazardous atmospheric conditions
- Excessive noise
- Radiation exposure
- Inadequate illumination or ventilation

There are many other situations or conditions that constitute an unsafe condition.

EXAMPLES OF UNSAFE CONDITIONS

- The removal of guards is considered an unsafe act which, in turn, creates an unsafe condition.
- Failure to warn people of danger is considered an unsafe act which, in turn, creates an unsafe condition.
- Drinking alcohol or the use of drugs is an unsafe act which could lead to creating an unsafe condition.
- Improper placement of material is considered an unsafe act which becomes an unsafe condition.
- Horseplay is indeed an unsafe act which usually terminates in an unsafe condition.
- Operating equipment at improper speeds is an unsafe act which quickly becomes an unsafe condition.

UNSAFE ACTS

The definition of an unsafe act, according to The American Society of Safety Engineers is

...a departure from an accepted, normal, or correct procedure or practice which has in the past actually produced injury or property damage or has the potential for producing such a loss in the future; an unnecessary exposure to a hazard; or conduct reducing the degree of safety normally present. Not every unsafe act produces an injury or loss.

By definition, all unsafe acts have the potential for producing future accident injuries or losses, an unsafe act may be an act of commission (doing something which is unsafe) or an act of omission (failing to do something that should have been done).

Unsafe acts are

- Operating without authority
- Failure to warn or secure
- Operating at improper speed
- Improper lifting
- Servicing equipment in motion
- Horseplay
- Drinking alcohol or drug intake
- Failure to wear protective equipment
- Using defective equipment
- Removing safety guards

There are many other practices that are considered to be unsafe acts.

EXAMPLES OF UNSAFE ACTS

- Improper ventilation is considered an unsafe condition, but failure to correct the situation is considered an unsafe act.
- Excessive noise is considered an unsafe condition, but failure to sound dampen or provide hearing protection is an unsafe act.
- Poor illumination is considered an unsafe condition, but failure to provide better lighting is considered an unsafe act.
- Defective tools and equipment is considered an unsafe condition, but failure to fix tools and equipment is an unsafe act.
- Poor housekeeping is an unsafe condition, but failure to correct the condition is considered an unsafe act.
- Refusing to purchase required safety equipment is creating an unsafe condition, but the act of refusing is considered an unsafe act.

ACTS OF GOD

- Snow storms
- Lightning
- Floods
- High winds
- High tides
- Rains

POOR OUTLOOK

You can spend great sums of money to correct unsafe conditions, but when the workers do not watch what they are doing, they are inviting an accident to happen.

We all know the four horsemen of safety, and they are

- Apathy — lack of concern or indifference
- Complacency — content or lack of vigilance
- Distraction — inattention or absent-mindedness
- Deviation — taking shortcuts or willfulness

Several years ago, I was asked to overlook a safety violation, a condition so hazardous that an accident was sure to happen to that person. I replied, "Asking me to overlook a safety violation is asking me to compromise my entire attitude towards the value of your life." Needless to say, the job was stopped for an hour in order to allow maintenance workers time to correct the unsafe condition. The one thing that impressed me the most during that hour was the employee's change of attitude, from anger to acknowledging that the condition was indeed hazardous.

When you notice a steady increase in your injury rate, take a walk to the work-sites and take a good and long look. It is certain that you will see many unsafe conditions and unsafe acts.

WHAT TO DO ABOUT AN UNSAFE CONDITION

1. Remove hazard if possible.
2. If danger cannot be removed, guard the people from danger and notify the proper authority.
3. If guarding is not feasible, warn employees that an unsafe condition exists by posting a danger tag/sign and barrier tape. Also, if the danger is great, the area or danger should be isolated from workers.
4. Stay on top of actions to correct unsafe conditions. If the condition is not corrected after a reasonable length of time, notify the proper authority or person(s) responsible.

WHAT TO DO ABOUT AN UNSAFE ACT

1. Stop the act.
2. Investigate why the unsafe act was committed.
3. Instruct employees in correct methods.
4. Train employees in the proper and safe methods.
5. Discipline employee only as a last resort and only when the employee refuses to obey safety rules.

CORRECTING UNSAFE CONDITIONS

You should establish a priority list with the most hazardous conditions first, and then on down to the least hazardous. Start with the cause of the most serious injuries, especially those that may lead to disaster, loss of lives and heavy property damage. Serious health hazards and other major concerns must be corrected promptly.

SUMMARY

When you properly assign the accident cause to an unsafe condition or to an unsafe act, you will find the reason why an accident happened. Unless you cannot tell the difference, you will always wonder what went wrong. Try this on your next accident investigation and you will find each time what caused the accident, and how it could have been prevented.

If the cause is an unsafe condition, you will want immediate attention to correct it. On the other hand, you may discover that the accident was caused by an unsafe act, then it is possible that the employee may need retraining or a warning. Whatever you do, try to properly assign the accident cause. You may bounce back and forth from one of two causes, but in the end, you will know exactly what caused the accident and how to prevent a repeat. And remember, an accident that has occurred is only the symptom of some underlying cause.

4 SAFETY INSPECTION

"You get what you inspect and not what you expect."

Unknown

LOOKING FOR HAZARDS

Do not wait until you have suffered a number of accidents to begin inspecting your work-sites. An accident, or worse, an injury, is very expensive and demoralizing to your productive effort. This brings to mind a story about a farmer who used to go over his field the day before plowing and kill all the rattlesnakes he could find. When asked if it was a dangerous thing to do, he replied, "I reckon so, but if I don't kill 'em off today when I'm looking, they might get me tomorrow when I'm not looking."

There is much logic in that farmer's thinking. In fact, his inspection of the field reflects a basic principle of accident avoidance through safety inspections. In other words, remove the hazard *before* it causes an accident. The next time you conduct a safety inspection, remember that an unsafe condition much like the farmer's rattlesnakes can get your workers when they least expect it to. So take a tip from the farmer and inspect your work-sites and remove all the hazards before they cause an accident.

DETECTING UNSAFE CONDITIONS AND UNSAFE ACTS

During your inspection, you should not only look for unsafe conditions, but you should observe unsafe acts. Some unsafe acts are easy to detect, especially when you have a written work procedure, and when the employees have been properly trained to do their work.

On the other hand, the process may be new and untried. In either case, you should involve the workers in trying to find a better and safer way to do the job. You should also investigate all accidents and near-misses which warn that a process is not working right. Unsafe acts, when detected, can be corrected by training, which I will explain in Chapter 7.

A SUPERVISOR'S RESPONSIBILITY

An experienced supervisor will not have any trouble eliminating unsafe conditions or unsafe acts. Most supervisors have a good knowledge of the equipment, especially when it comes to getting more work done without sacrificing quality or safety. During each step of the operation, the supervisor should reflect on this question, "What could go wrong and cause an accident?" For example, if a worker is removing metal chips from a drill press while the machine is running you should consider if there is a better way to remove the chips.

CHECK EACH STEP OF A JOB

When looking for hazards, check each step of a job. Talk with the workers often; they can be very helpful. And if the workers' ideas are sound, recognize this by providing an incentive award. The idea may not only be safer, but may also save money as a result of less accidents, which also means less downtime.

You should question the workers regarding near-misses. They are a preview of an accident, or worse, an injury. In other words, do not discount the possibility of a serious injury from a known hazard just because an accident is yet to occur.

Another way to look for unsafe conditions and unsafe acts is to study previous accidents at the same work-site. This helps to pinpoint the areas where accidents occur. Your personnel department, or your safety office can help you to identify those areas where injuries are most likely to happen, especially when statistics are maintained.

CONSTRUCTING A SAFETY INSPECTION LIST

Supervisors should construct their own safety inspection list of unsafe conditions and unsafe acts. I will not attempt to construct a safety inspection list because all work-sites are different and a typical safety inspection list would not be appropriate for all operations. I will, however, list one unsafe condition and one unsafe act that I believe apply to all work-sites, and the rest is up to you.

Safety Inspection List

Unsafe conditions	S = Satisfactory	U = Unsatisfactory
Aisles and walkways	_____	
Storage of materials	_____	
Scraps and debris	_____	
Unsafe acts		
Improper lifting	_____	
Smoking where prohibited	_____	
Removing safety guards	_____	

IMMINENT DANGER

If an imminent danger situation is discovered during your safety tour, you should take appropriate action to stop the job and warn personnel of the danger. It may be necessary to evacuate the area. The definition of imminent danger is, "A hazardous situation which may cause death, loss of a facility, severe injury, severe occupational illness or major property loss, and it is likely to occur immediately or in a short period of time."

IN GOD WE TRUST

Francis Bacon, in the 16th century, once said, "Wise distrust and constant watchfulness are the parents of safety." His statement of wise distrust means, do not believe everything you are told. Rather, go out into the workplace and look. It does not mean that your employees are lying, it may just be that they do not know, or, they believe no one can do anything to correct the hazards. Bacon's other statement regarding constant watchfulness does not mean baby-sitting the work-site. It means you should know your operation and your employees to the point that you will know how to prevent any process from going out of control.

When your employees see you inspecting and asking questions, they will know that you are serious about safety, and in this regard, they will not hesitate to report unsafe acts or unsafe conditions to their supervisors.

MEANINGFUL SAFETY INSPECTION

To be effective, the safety inspection process has to be a daily routine and not something that is done only when things go wrong. Supervisors must be constantly alert to the condition of their work-sites, so as to detect and correct unsafe conditions. Also, they should look to see that the workers are employing safe working practices. When the supervisor does not care, the workers may develop poor work habits that lead to accidents and injuries. When a supervisor continuously shows his or her interest by being firm and by taking immediate corrective action whenever an unsafe act is observed, the employees will soon realize that they are expected to work safely at all times, and in this way, they will develop good working habits.

GOOD HOUSEKEEPING

There are three good reasons why you should maintain good housekeeping throughout your work-sites:

1. Good housekeeping contributes to the elimination of accidents and fires.
2. Good housekeeping helps conserve space, time, material and labor.

3. Good housekeeping contributes to high morale. No one really wants to work in a pig sty.

When everybody does their share to maintain a clean work-site, all will benefit by having a safe place to work.

Good housekeeping means more than one thing, it also means that walking and working surfaces are free from tripping, slipping, and stumbling hazards. It also means that all work-sites are free from clutter. Work-sites include both indoor and outdoor situations, especially those of construction projects where clutter such as a board with a nail can bite through the shoe of a worker.

SAFETY INSPECTORS

Some companies use safety committees or safety coordinators to inspect the workplace. When they do this, it is in support of the supervisors daily inspection. Each worker is also responsible to inspect his or her work-site. They should do this every day and immediately report hazards or defects that they cannot fix to their supervisors. Maintenance workers should also inspect for safety when they are performing their checks of equipment. The tool room employees should inspect all tools to see that they are kept in safe condition.

When everyone is involved in the inspection process, good attitudes toward safety are enhanced to the point that accidents and injuries are practically eliminated, and in that respect, you will have avoided an OSHA inspection because of your low accident rate.

Remember, your employees have a right *by law* to a safe and healthful environment in a workplace that is free of recognized hazards. The OSHA Act imposes a legal liability on you to conform to certain standards.

SUMMARY

Take your time during a safety inspection; look at every part of the operation. I am not saying that you should be involved each time, but you should know if your workplace is unsafe. Remember, the hazards found during an inspection are directly proportional to the time spent looking for hazards. If your people do not spend enough time looking, they will not find many hazards. The easy ones will most likely be the first hazard they will see.

As you get deeper into the inspection process, your skills will sharpen, and you and your people will see situations that will eventually become serious hazards. Make sure that your walking and work areas are free of oil, grease, water, ice, and other slipping hazards. It is also a good idea to take a new worker along during the inspection tour. Supervisors who have done this all say that it is a good way to introduce a new worker to safety.

5   ACCIDENT INVESTIGATION

"Facts do not cease to exist because they are ignored."

Aldous Huxley

THE DEFINITION OF AN ACCIDENT

An accident can be defined in many ways, but they all mean the same thing. Consider the following three definitions and the effects they have on the order of good business:

- An accident is an undesired event that results in physical harm to a person or damage to property.
- An accident is an unwanted interruption of a desired course of action.
- An accident is an unplanned, unforeseen, unwanted occurrence that interrupts or interferes with the orderly progress of an activity.

Webster's Dictionary defines an accident as (1) an undesirable or unfortunate happening, casualty, or mishap, and (2) anything that happens unexpectedly without plan, or by chance. This definition is not completely acceptable to some safety professionals and industrial managers. Some accidents are planned — unintentionally, but, nevertheless, planned. For instance, using a rickety ladder or operating a machine without a guard are good examples of conscious decisions to commit an unsafe act, which can result in accidents.

CHASING THE ACCIDENT THREAD

Probably the best way I can explain this is to repeat information from an article that I wrote several years ago for *Lifeline* magazine, and which was subsequently reprinted in *The Safe Foreman* publication in November of 1984. The article is titled, "Dig a Little Deeper".

Lots of people think that we only need to investigate mishaps or accidents when an unsafe act or an unsafe condition is found. This kind of logic seems adequate, but quite the opposite is true. Shallow investigations do not reveal all the casual factors surrounding a mishap. Consequently, the mishaps go on and on because we have not uncovered all the factors "why." Mishaps and near-mishaps should be investigated thoroughly to determine root causes.

Kenneth R. Andrews of The Harvard School of Business Administration said in 1973, "Every mishap, no matter how minor, is a failure of organization." Chasing the thread of a mishap requires care, attention, and perseverance. Investigators have a tendency to stop digging when certain facts or causes are only the tip of the iceberg.

Let's look at a typical industrial mishap involving a maintenance worker falling off a 6-foot stepladder. The initial investigation reveals that the worker was standing on the topmost portion of the ladder while attempting to replace a burned-out light bulb. Standing on the top part of the ladder is considered an unsafe act. Do we stop our investigation and blame the worker? Continuing our investigation, we find that there are no other stepladders in the maintenance shop except 6-foot stepladders. So it appears that management is at fault for creating the unsafe condition.

As we continue to chase the thread, we find that the worker was never trained in ladder safety. Digging deeper, we find that this department has had mishaps of this type before, but no action was ever taken to order a longer stepladder. Why? Perhaps no one thought it was important or that it was the worker's fault. Should we stop here? No way! We want to know if a longer ladder was ever requested but dismissed because of cost. We want to look into the company's training program, safety meetings, and safety inspections.

By chasing the mishap thread as far as we can go, we may find that the worker was properly trained, or that the company had a longer stepladder in another department and that the worker created his own unsafe condition, but we will not know this until we have explored all the obvious and not-so-obvious casual factors. To do our investigation any differently invites near-miss situations that lead to minor events, and then on to more serious or catastrophic mishaps.

The above situation is relatively basic, but if we do not investigate all mishaps properly, we will not be able to investigate complex mishaps involving loss of personnel and property. We should learn from our mistakes and recognize that near-miss or almost mishaps are only a preview of coming events; and with that said, I am reminded of a saying by Haliburton, "Hear one side, and you will be in the dark; hear both sides, and all will be clear."

THE GREAT MYTH

How many times have you heard people say after an accident, "that's the price of progress", or "when your number is up, you go", "that's an act of God", or "the law of averages says it was bound to happen". These are irresponsible beliefs and unquestionable obstacles to the order of good business.

Also, it is a myth that some people lead charmed lives, just as it is a myth that accidents always happen to the other person. In reality, we are all exposed to some sort of hazards. Our world is made up of so many hazardous conditions that the only way we can be safe is to recognize and know the hazards inherent in each job or situation encountered. Then, and only then, can we adequately protect ourselves. An unknown person once said, "I wonder why somebody didn't do something, then I realized that I was somebody."

THE BENEFITS OF ACCIDENT INVESTIGATION

The purpose of an accident investigation is to find the causes of the accident in order that appropriate preventive measures can be taken. Briefly, the reasons to investigate are as follows:

- To learn why the accident happened.
- To make changes that will prevent a repeat accident.
- To make employees aware of the hazard.

Basically, the investigation must answer the following questions:

- *Who* was injured?
- *When* did the accident occur?
- *Where* did the accident occur?
- *Why* did the accident occur?
- *How* can it be prevented?
- *What* type of weather, lighting, temperature, noise, etc. was involved?

Getting to the scene of an accident promptly is very important. The reason for this is that after the accident has happened and time passes by, it becomes more difficult to obtain facts. Conditions change rapidly and people forget. Again, prompt accident investigation is a must in order to ensure that the accident will not be repeated.

Accidents are rarely caused by a single factor, but instead, by several conditions or events that come together at the same time. In some organizations, close-shaves, near-misses and incidents are investigated with the same spirit of inquiry as that of major accidents.

HOW TO INVESTIGATE

The investigator should be equipped with the following:

- Camera and film
- Clipboard, paper and pencil
- Magnifying glass (5× or 10×)

- Report forms
- Strong gloves
- High visibility barrier tape
- Cassette recorder and tapes
- Graph paper
- Ruler and tape measure
- Identification tags for parts
- Transparent tape
- Specimen container
- Compass
- Flashlight

Priorities at the Mishap Scene

1. Arrive safely and take charge
2. Observe the overall scene upon arrival and evaluate the situation
3. Care for the injured
4. Protect others from injury
5. Remove onlookers from the immediate area

Second Set of Priorities at the Mishap Scene

1. Preserve evidence
2. Protect the mishap site
3. Secure the evidence
4. Keep the boss informed

Preservation of Evidence

- Control crowds and traffic
- Take charge
- Take photos or sketches
- Hold witnesses together
- Erect barriers

Gathering the Evidence

- Gather samples of evidence (oil, glass, metal, etc.)
- Label samples
- Take measurements
- Identify photos and sketches
- Identify witnesses by name, address, etc.

Interviewing

- Do the interviewing in a quiet, neutral non-threatening location.
- Tell the witness the purpose of the interview.
- Do not spring any surprises about the nature of the interview.
- Take down essential information.
- Let the witness tell the story in his or her terms, not yours.
- Do not interrupt the witness.
- Encourage the use of sketches.
- Use a cassette recorder, however, get his/her approval first.
- Take notes.
- Avoid leading questions. Ask the witness to rephrase when necessary.
- Use tact and diplomacy; be neutral.
- At the end of the interview, ask if he/she knows anyone else who saw the event.
- After the interviews, reconstruct new sketches, diagrams and/or charts.

SUMMARY

An accident is unfortunate, costly, and it can cause untold suffering. The purpose of investigating accidents is to uncover all the details regarding how and why the accident happened. Without accident investigation, you are bound to have the accident be repeated many times without knowing the cause or how the same accident can be prevented.

Investigating accidents will tell you whether the accident was caused by an unsafe act, or by an unsafe condition, or both.

You should also investigate the close calls (near-miss), because they preview an accident waiting to happen.

The law books are full of situations where accidents were not investigated and top management was hauled into court and made to pay huge sums of money for their neglect. It makes good business sense to investigate all accidents, no matter how seemingly minor they are.

6   JOB HAZARD ANALYSIS

"Knowledge exists to be imparted."

Emerson

ANALYZING SAFETY AND HEALTH HAZARDS

Many years ago when I decided to become a safety professional, job hazard analysis was called job safety analysis. That was before the Occupational Safety and Health Act of 1970, which brought about many changes. More specifically, the act brought safety and health together.

The newly formed Occupational Safety and Health Administration (OSHA) under the Department of Labor, recognized that health hazards needed to be analyzed much the same way as safety hazards. Therefore, an OSHA booklet, number 3071, titled "Job Hazard Analysis" was published in order to provide industry with one method to analyze both safety and health hazards.

I will not attempt to write something better than a process that already works well. Instead I have utilized the OSHA publication and have added some thoughts of my own.

BENEFITS OF JOB HAZARD ANALYSIS

As a manager, you should take action to isolate the hazard from your people, or isolate your people from the hazard; and with that said, I will explain what benefits you will derive by doing this.

Job related injuries occur every day in the workplace. Often these injuries occur because employees are not trained to follow the proper procedures. One way to prevent workplace injuries is to establish safe job procedures and then train employees to work following that procedure.

Establishing safe job procedures is one of the benefits of conducting a job hazard analysis; carefully studying and recording each step of a job, identifying existing or potential job hazards (both safety and health), and determining the best way to perform the job to reduce or eliminate these hazards.

Improved job methods can reduce costs resulting in employee absenteeism and workers' compensation, and can often lead to increased productivity. The Occupational Safety and Health Administration confirms this fact in their booklet.

The following pages will contain all the information you will need to understand job hazard analysis, and will provide guidelines which will allow your people to conduct a step-by-step analysis.

SELECTING JOBS FOR ANALYSIS

A job hazard analysis can be performed for all jobs in your workplace, whether the job is "special" (non-routine) or routine. Even one-step jobs, such as those in which only a start and stop button is pressed can, and perhaps should be, analyzed by evaluating surrounding work conditions.

To determine which jobs should be analyzed, first review your job injury and illness log. Obviously, a job hazard analysis should be conducted first for jobs with the highest rates of accidents and disabling injuries. Also, jobs where "close calls" have occurred should be given priority. Analyses of new jobs and jobs where changes have been made in processes and procedures should follow. Eventually, a job hazard analysis upon completion should be made available to your employees for all the jobs in your workplace.

INVOLVING THE EMPLOYEE

Once you have selected a job for analysis, discuss the procedure with the employee performing the job and explain its purpose. Point out that you are studying the job itself, not checking up on the employee's job performance. Involve the employee in all phases of the analysis, from reviewing the job steps to discussing potential hazards and recommended solutions. You should also talk to all other workers who have performed the job.

CONDUCTING THE JOB HAZARD ANALYSIS

Before actually beginning the job hazard analysis, take a look at the general conditions under which the job is performed and develop a check list. Below are some sample questions you might ask.

- Are there materials on the floor that could trip a worker?
- Is lighting adequate?
- Are there any live electrical hazards at the job site?
- Are there any explosive hazards associated with the job or likely to develop?
- Are tools, including hand tools, machines, and equipment in need of repair?
- Is there excessive noise in the work area, hindering worker communication?

- Is fire protection equipment readily accessible and have employees been trained to use it?
- Are emergency exits clearly marked?
- Are trucks or motorized vehicles properly equipped with brakes, overhead guards, back-up signals, horns, steering gear, and identification as necessary?
- Are all employees who are operating vehicles and equipment properly trained and authorized?
- Are employees wearing proper personal protective equipment for the jobs they are performing?
- Have any employees complained of headaches, breathing problems, dizziness, or strong odors?
- Is ventilation adequate, especially in confined spaces?
- Have tests been made for oxygen deficiency and toxic fumes?

Naturally, this is by no means a complete list because each work-site has its own requirements and environmental conditions. You should add your own questions to the list. You might take photographs of the workplace. If appropriate, for use in making a more detailed analysis of the work environment.

IDENTIFY EACH STEP AND HAZARD

Nearly every job can be broken down into steps. In the first part of the job hazard analysis, list each step of the job in order of occurrence as you watch the employee performing the job. Be sure to record enough information to describe each job action, but do not make the breakdown too detailed. Later, go over the job steps with the employee. After you have recorded the job steps, next examine each step to determine if a hazard does exist or that it could happen.

Ask yourself these kind of questions:

- Is the worker wearing protective clothing and equipment, including safety belts or harnesses that are appropriate for the job?
- Are work positions, machinery, pits or holes, and hazardous operations adequately guarded?
- Are lock-out procedures used for machinery deactivated during maintenance procedures?
- Is the worker wearing clothing or jewelry that could get caught in the machinery?
- Are there fixed objects that may cause injury, such as sharp machine edges?
- Are the workers, at any time, in an off-balance position?
- Can a worker fall from one level to another?
- Can a worker come in contact with live exposed electrical parts?
- Can the worker be injured from lifting or pulling objects, or from carrying heavy objects?
- Do environmental hazards — dust, chemicals, radiation, welding rays, heat, or excessive noise — result from the performance of the job?

These are but a few of the questions you should ask yourself and the worker who knows the job better than anyone else.

RECOMMENDING SAFE PROCEDURES AND PROTECTION

After you have listed each hazard or potential hazard and have reviewed them with the employee performing the job, determine whether or not the job could be performed in another way to eliminate the hazards. If safer and better job steps can be used, list each new step, such as describing a new method for disposing of material. List exactly what the worker needs to know in order to perform the job using a new method. Do not make general statements about a procedure, such as "be careful".

You may wish to set up a training program using the job hazard analysis in order to train your employees to use the new procedure, especially if they are working with highly toxic substances or in dangerous situations and environments. If no new procedure can be developed, determine whether any physical changes, such as redesigning equipment, changing tools, or adding machine guards, personal protective equipment or ventilation, will eliminate or reduce the danger.

Go over the recommendations with *all* employees performing the job. Their ideas about the hazards and proposed recommendations may be valuable. Be sure that they understand what they are required to do and the reasons for the changes in the job procedure.

REVISING THE JOB HAZARD ANALYSIS

Remember, a job hazard analysis can do much toward reducing accidents and injuries in your workplace, but it is only effective if it is reviewed and updated periodically. Even if no changes have been made in a job, hazards that were missed in an earlier analysis could be detected.

If an accident or injury occurs on a specific job, the job hazard analysis should be reviewed immediately to determine whether changes are needed in the job procedure. This is also a good subject to discuss during stand-up safety meetings.

A job hazard analysis also can be used to effectively train new employees on job steps and job hazards. Finally, the job hazard analysis will save you money and increase productivity.

SUMMARY

Performing a job hazard analysis will tell you everything you need to know about a certain job. It will also help you understand how and why accidents happen. Job hazard analysis can be a strong motivator during training sessions, especially with new employees. The analysis is also very useful when you decide to retool, install new machinery, or make changes to a process.

Job hazard analysis will save you money and will increase your profits. It demonstrates to the employees that management cares about their well being. They can see that they are part of the work process. Job hazard analysis provides the worker with an opportunity to get directly involved with safety.

7   SAFETY TRAINING

"Skill to do comes of doing."

Emerson

IMPORTANCE OF SAFETY TRAINING

People are not born knowing everything — they must be trained for each new job. Oftentimes, it is taken for granted that the new employee has received prior job and safety training from a previous employer. That assumption could be disastrous for your company and the new employee.

You should take the position that your new employee, regardless of prior training, must be trained in safety prior to being assigned to the new job. The new employee will most likely tell you that he or she has received prior training and that more training is a waste of time. When this happens, ask yourself this question: "Do I really know for sure that the new employee has received the required safety training?" Besides, your job site, tools, and machinery will most likely be different, and the new employee may be facing new and unfamiliar hazards.

PLEASING THE BOSS

Years ago, I was assigned to do a job that I had no knowledge of, but I was willing to give it a try ... to please the boss. Consequently, I was injured in the process of doing the job. You guessed it! I was not given any instructions whatsoever regarding the hazards of the job. I was laid up for quite a while, but I was fortunate that the company paid the medical costs. When I recuperated and returned to work, I was told that I was careless and was immediately given a termination notice and told that I was no longer needed.

Later on, I heard that several other workers were injured doing the same job. The company stayed in business 4 years and then failed. I do not know if the company failed because of unsafe conditions, but it is possible that they mismanaged every segment of their business the same way.

AN OBLIGATION

The employer has a responsibility to instruct all employees in the hazards of the work. To merely say, "Here's the job, do it safely", is not enough. There is a lot more to getting new workers on the right path than the all too familiar cop-out of handing the workers a piece of paper with some instructions that, in all probability, they will not read or might not understand.

When you have performed job hazard analysis for each job as outlined in Chapter 6, use this information to train your new workers. You can also use the analysis to correct an employee with a poor safety record. You will not find a better lesson plan, and it will be believed by your workers.

Training is also a requirement of the OSHA Act. You are required *by law* to train your employees in the safe ways to work. Do not get in hot water with OSHA by skipping the training; it can cost you money.

TRAINING YOUR EMPLOYEES

You should consider the simple but effective tell-show, test-check method which provides a systematic and reliable replacement for the old hit-or-miss approaches.

- **Tell** — instruct clearly and completely with a step-by-step approach.
- **Show** — Demonstrate the job or operation and stress key points.
- **Test** — Ask questions to see if your instructions are understood.
- **Check** — Follow-up at the job site with frequent visits.

An old Chinese proverb says it well, "I hear and I forget, I see and I remember, I do and I understand."

AN ON-GOING PROCESS

You should advise your instructors to keep the lessons simple and to the point. They should always leave the employee with the feeling that more information is available and tell them where to get it.

Do not expect your supervisors to do the training. It does not work, because the employee should go to the supervisor with their questions about the training they have received. That gives them assurance of knowing whether or not they have received proper training. Remember that no one is everything to everyone. We all have our jobs. A supervisor provides supervision while an instructor trains workers to do the job the right way and in a safe manner. It is recognized that there are two other ways to train employees beside a formal instructor training program.

On-The-Job Training

Oftentimes, the worker learns by watching and working with an older employee; one who has a good track record of producing a quality product in a timely and safe manner. This type of training does not replace the formal instructor training, but instead puts into practice what has been taught in a classroom.

You should never assign a new worker to a mediocre worker for training purposes. This invites trouble. The new worker will learn things the wrong way and will invariably give you problems regarding the quality, safety, and timeliness of your product. Once bad habits set in, it is very difficult to correct an employee who has been improperly trained, not to mention the cost.

On-the-job training can be very beneficial because the supervisor can observe the training and offer suggestions. The supervisor can also see if the new worker brings any bad habits or a poor attitude to your company.

Stand-Up Meetings

This method of training can be very effective, or it can cause problems. First, let me say how it can go wrong. Once a week, the supervisors gather their people together to give them a safety talk. At construction sites this is called a "tail-gate talk". Some supervisors who should never have been selected in the first place to be supervisors will demean safety.

I remember in my many years in the construction and manufacturing business, coming across supervisors who started their presentations by saying, "This will only take a minute." This is like saying, "We are wasting time, but we have to talk safety."

A supervisor should *never* apologize for safety. When they do, it is an invitation to trouble that will have a "domino" effect throughout the group. It will also affect other controls such as quality, absenteeism, etc. On the other hand, the stand-up safety meeting, when properly administered, will become a strong motivator.

Supervisors should involve their subordinates in the safety discussions. They know where the problems are, and in most cases, have the capability to help solve the problems. The supervisor should encourage feedback of information from the working level. It has to be a two-way street in order to communicate safety.

SUPERVISORS' RESPONSIBILITY

It is the supervisors' responsibility to be knowledgeable about safety and to develop in their workers a proper attitude toward safety. Safety is a state of mind and if you do not keep it active, the employees will lose sight of it.

However, by giving good safety talks, the supervisor will educate and motivate the whole team to observe safe work practices.

The supervisor may want to discuss a recent injury suffered by a worker, or to talk about wearing personal protective equipment. Also, the importance of performing each job safely could be stressed by demonstrating how it should be done.

A big plus is letting your employees participate in the talks, rather than forcing them to just stand and listen to the preaching. Showing films is another way to inform the workers. The main thing is whatever the supervisors say at the meetings, they should not demean safety; your employees deserve better.

You should sit in and listen to the talks during a meeting, you may be surprised. You may be able to offer a better way to communicate safety. Thomas Edison put it very well when he said, "there must be a better way — find it!"

Often we hear people say, "Experience teaches". That is misleading, for experience teaches nothing. Only people teach, not experience.

SUMMARY

Very few become outstanding workers by being careless or indifferent to safety. A firefighter would never attempt to go into a burning building without protective gear. Nor would they do so without instructions and drills. The same should be true for your employees. They should be trained to do the job, to recognize the hazards, and to do it the proper, safe way each time.

Training the employee to work safely is not separate from job skill training. As a matter of fact, they are one and the same and therefore, should be given at the same time. When the training is done this way, you will reduce accidents and injuries and your profits are bound to increase. Make certain that your supervisors support your safety policy, and when they do not, take appropriate corrective action to ensure that they will in the future.

8   HUMAN FACTORS

"He that will not apply new remedies must expect new evils."

Francis Bacon

THE STUDY OF HUMAN FACTORS

Human factors engineering is also called ergonomics. The word ergonomics is taken from the Greeks: ergo is the Greek word for *work;* nomos is the Greek word for *knowledge.* Therefore, the word ergonomics means work knowledge.

Ergonomics has been defined as the study of the anatomical, psychological, and physiological aspects of a human being in a working environment. Its object is to optimize human safety, health, comfort, and efficiency. This sounds complicated, but it really is not.

In other words, ergonomics or human factors engineering can maximize the worker to increase productivity, job satisfaction, and job attitude. When this happens, injuries will be minimized and hazards will be practically eliminated.

PEOPLE ARE IMPORTANT

The most important component in any work application is the people who are directly involved in the task. Most tasks require only basic adjustments in order to improve working conditions. However, there are some specialized work groups that require greater attention to ensure maximum productivity, and at the same time, a greater assurance of safe conditions. A good example is the specialized group of workers who handle chemicals. The work-site modification for the specialized groups are often achieved at minimal cost and will benefit other groups of workers as well. In fact, the entire company may benefit from the special effort.

PURPOSE OF HUMAN FACTORS

The major goal of human factors engineering is to design safety in the job, in the machine, and in the environment. It does not try to make people perform in ways other than "what comes naturally". The major concerns are

- Engineering and producing equipment for the population that will use it.
- Designing a system so that machines, human tasks, and the environment are compatible with the capabilities and limitation of people.
- Designing the system to fit the characteristics of people rather than retrofitting people into the system.

HUMAN LIMITATIONS

The intent is to ensure that the work, the equipment, and the environment are so designed or modified as to fit the job to the person rather than the person to the job. Consider the following human differences:

- Women's legs are longer than men's, but their arms are shorter.
- Women have 20% less strength than men.
- People working upright are prone to job stress.
- Working with outreached arms can result in back pains.

These are but a few of many human limitations that when not recognized can cause loss of productivity, accidents, material loss, and human suffering. A good point to remember is, "its easier to change the workplace than to change or train people."

STUDY THE WORKER

Ergonomics is concerned with the study of people in their work environments, the tools used, and the machinery required to produce a product. Therefore, it is necessary to examine the physiology of work and fatigue, the physical structure of the person, the tool design, the work layout, the task, and the environmental hazards of the operation.

Thus, we have the man-machine-environment system. Fitting the worker to the job and workplace in the most cost-effective manner requires analysis of human physical and psychological needs and adjustment or redesigning of tasks and tools to account for individual differences.

It begins simply by observing the task, the worker, and the work environment, then analyzing each movement or position particular to the task. Proper lighting, rest periods, noise, temperature, and ventilation must be assessed because they all contribute to worker fatigue.

THE HANDICAPPED WORKER

Perhaps one of the best examples of fitting the task and work-site to the capabilities of the worker may be found in the area of vocational rehabilitation and opportunities for the handicapped. Studies have shown that industry benefits when work stations are redesigned in order to permit a worker to return to a job after an injury, or to allow a handicapped individual to become an active worker.

Expenditures for modifications are often less costly than replacing an experienced worker or incurring disability payments and costly accident insurance premiums. Work-site modifications not only reduce injuries, but also boost morale. This often leads to better awareness of safety.

FUTURE RISK

The advent of the video display terminals (VDTs) is bringing some concerns to the Occupational Safety and Health Administration (OSHA). Notable safety and health groups are often asked about possible safety and health problems associated with VDTs.

Some concerns include high voltage electricity, noise, radiation, and birth defects. Currently, OSHA has no reliable information that any birth defect has ever resulted from a pregnant woman working at a VDT. However, national agencies are still studying the potential problem (OSHA has a number of electrical requirements applicable to VDTs).

The equipment must be properly installed, used, and grounded to ensure employee safety. The noise levels should be kept within comfortable limits. Loud sounds that are unacceptable should be shielded by sound absorbent screens or hoods or placed in a separate room. Absorbent materials such as acoustical ceiling tiles, carpets, curtains, and upholstery can reduce noise. National health agencies and the National Institute for Occupational Safety and Health (NIOSH), and others, have measured radiation emitted by VDTs. The tests show that levels for all types of radiation are below those allowed in current standards.

Video display operators do sometimes report eye fatigue and irritation, blurred vision, headaches, dizziness, and pain or stiffness in the neck. They also experience problems with shoulders, back, arms, wrists, and hands. These problems usually can be corrected by adjusting the physical and environmental setting where the VDT users work. The relation of the operator to the keyboard and the screen, the operator's posture, the lighting and the background noise should be carefully examined to prevent discomfort.

WORKER COMFORT

The decision whether to have a worker stand or sit at a work station is influenced by several factors:

Standing
- Greater movability
- Greater use of work station

Sitting
- Less loading postural muscles
- Less freedom of movement

Therefore, it appears that sitting is less stressful to the worker than standing, however, poor design of the work seat may actually increase static loading of some muscles. For example: when the workbench is too high, the elbows will be held away from the body, thereby causing undue fatigue.

OTHER HUMAN PROBLEMS

Humans, not unlike machines, have certain capabilities and limitations, and these should be recognized for greater work efficiency. Consider some of the problems that you encounter every day: poor attitudes, behavior, nervousness, anger, alcoholism, fatigue, plus many more human quirks that inhibit production. For example, a large person cannot work in a cramped space; some people can lift great weights while others cannot.

Human beings are often required to perform innumerable tasks as part of their daily working situations. Not unlike any machine when it becomes overloaded, the human machine can break down; it can suffer temporary or even permanent damage. In the event that a task demands more of the workers than they can sustain, rest breaks should be provided. It has been found that taking a short pause every 30 minutes has increased productivity when compared with no breaks.

Human factors should always be considered in job assignments, and when this is done, the results will be increased efficiency, a decrease in human error, and a significant reduction in accidents and injuries. Remember, machines have built-in upper tolerance limits. Humans, on the other hand, do not. They need to be challenged, but not overburdened to the point where they become susceptible to accidents and injuries. It is not my intent to make you an expert in the area of human factors engineering or ergonomics, but instead, to provide you with an understanding that will enable you to seek the expertise of a professional to assist you to preserve your human resources. You should look at human factors at the same time you perform a job hazard analysis, which is found in Chapter 6.

SUMMARY

You should design the employee work-sites or work stations in such a way that workers will become more productive without fatigue. Take a tour of the work areas and solicit ideas from the workers. When they offer suggestions, tell them that you will look into the problem and get back to them.

Look at the injury and sick leave records, then you will know where to look. If you are too busy, hire a consultant to look over the areas with the most problems. If you have a suggestion box, encourage your worker to submit problems or ideas. Most of the time, they know the answer before they decide to go the suggestion route. Believe it! You can improve your productivity and reduce your injury losses when you take a good look at human factors and how they affect your pocketbook.

9   FIRE PREVENTION

"Fire is the best of servants; but what a master!"

Carlyle

A RESPECT OF FIRE

Fire is a double-edged sword! It can perform wondrously, or it can destroy your business. If you doubt this, talk to your State Fire Marshal. Fire has to be respected or it will turn on you when you least expect it to. The respect of fire is the object of fire safety, and fire safety is the protection of life and property from the ravages of fire.

FIRE-SAFE CONSTRUCTION

Your building design and construction must take into account a wide range of fire safety features. This also includes the contents of your buildings and the water supply needed for fire extinguishment. Think about this, your building provides a great potential for fire because of the large number of people working in or around the buildings. Your people could also bring about a careless or malicious act that could result in a fire.

The mechanical or electrical equipment needed to operate machinery and support systems are potential fire hazards due to faulty design, construction, or poor installation. Accumulations of combustibles waiting for disposal or in storage can provide the means by which fire could spread. Many fires disastrous to people and to property have occurred and will continue to occur unless proper consideration to the threat of fire is taken.

The fire safety of your company's buildings will depend on what is done to prevent a fire from starting. It will next depend on the type of construction of your building. And finally, your plan to minimize the spread of fire, should it happen.

FIRE PREVENTION

Good housekeeping is considered by top fire professionals as the major factor in fire prevention and control. Good housekeeping lessens the amount of material that can be ignited, and it also provides less materials that can be consumed if a fire breaks out.

Henry Ford was once asked what he would do if he was called upon to take charge of a business that failed. He replied, "No business I know ever went to the wall without accumulating a vast pile of dirt (trash), the dirt and all that goes with it — untidy thinking and methods — helped cause the failure. The first thing I would do would be to clean up that business."

Yes, there is nothing so wasteful of time, materials, and energy as disorder. It is also a fire waiting to happen. When your workers start to do something, and things are not in order (poor housekeeping), they immediately begin to waste time. It is a proven fact that your people can do the job quicker and more easily if the work-site, plant, or office is in order. Be very careful of what you store and what you store it in. Some users buy products for what they *can* do, not for what they are really intended for. Some products should not be stored together (see Chapter 11, Hazard Communication). The Material Safety Data Sheet (MSDS) for each chemical or substance you buy will state its fire reactivity and whether or not it is incompatible with other substances.

FIRE SAFETY CHECK LIST

Rubbish	Avoid accumulation
Flammables	Special storage — avoid spillage
Housekeeping	Neat and tidy work areas
Electricity	Grounded and good connections
Machinery	Clean and good maintenance
Smoking	Designated areas only*
Ventilation	Removal of fumes, dust, etc.
Extinguishers	Proper type and availability
Exits/Aisles	Clear, lighted and unobstructed
Evacuation Alarm	Periodically tested
Evacuation	Employees are instructed
Fire Bill	Lists responsibility of personnel

* If you want to know if your employees are smoking in non-smoking areas, look for cigarette butts on the floors of these areas. If you do find butts, you will know that your non-smoking policy is being ignored.

FIRE CAUSES

23%	Electrical (poor or improper installation)
18%	Smoking (violating no-smoking signs)
10%	Friction (improper machinery maintenance)
7%	Hot surfaces (burnables next to heat source)
7%	Open flames (welding, cutting, etc.)
5%	Flammable liquids (gasoline, alcohol, etc.)

FIRE CAUSES (continued)

5%	Portable heaters (gas furnaces, salamanders, etc.)
4%	Spontaneous (oily rags, improper storage, etc.)
3%	Arson (deliberately started fires)
2%	Mechanical sparks (foreign metal into machinery)
1%	Static electricity (sparks near flammable gases)
1%	Lightning (thunderstorms)

Note: The remainder are caused by other miscellaneous causes.

THE IMPORTANCE OF EXITS

Of the many factors involving your employees' safety, getting out of the burning building is considered the most important. Exits should be unobstructed and the surrounding areas well lighted, with lighted exit signs over the doors. Arrows should also point the way out of the building so that in the event of a fire or fire drill exits can be easily found.

It is recommended that every building or structure and every section or area in them should have at least two separate means of egress, so arranged that the possibility of any one fire blocking all of them is minimized. Exits should permit your workers to reach a place of safety before they are endangered by fire. The National Fire Protective Association, NFPA 101, Life Safety Code, provides a reasonable guide to exit requirements. OSHA, Part 1910.155, also provides fire protection guidelines for industry to follow.

OSHA also provides the scope and application regarding industrial fire brigades in Part 1910.156. A fire brigade is a small fire fighting unit within the industrial building that can respond to fires very quickly. Usually, the members have a collateral duty function. They are primarily industrial workers who can also respond to in-plant fires very quickly. Some workers also serve as volunteers with local fire departments.

FIRE EXTINGUISHERS

Fire extinguishers are special pressurized devices that release chemicals or water that put out flames. Fire extinguishers, when used properly, can keep small fires from becoming bigger fires. They also help provide escape through a small fire. And finally, they are used to contain a small fire until the fire department arrives to take over.

There are four classes of fire as follows:

CLASS A   These are fueled by ordinary combustible materials such as wood, paper, rags, plastics and rubbish. *EXTINGUISHER* (CLASS "A" FIRES): Water extinguishers — cools and soaks burning materials.

CLASS B   These are fueled by flammable liquids, gases, oil, grease, paints, gasoline and thinners. *EXTINGUISHER* (Class "B" OR "C" FIRES): CO_2 extinguishers to smother flames, or dry chemical extinguishers to blanket and smother flames with powder.

CLASS C   These are fueled by live electrical wires or equipment such as motors, power tools and appliances. *EXTINGUISHER* (CLASS "A", "B" OR "C"): Multi-purpose dry chemical extinguishers smother flames like the Class "B" types, but with a different powder, or a liquified extinguisher which smothers and cools flames (CO_2). Also, the Halon 1211 extinguisher can be used.

CLASS D   These are fueled by combustible metals, magnesium, titanium, zirconium, etc. *EXTINGUISHER* (CLASS "D" ONLY): special dry-powdered agents.

SUMMARY

A fire is very destructive and can bring the very best company to its knees. Some companies never recover from fire disasters. No amount of insurance can protect against the loss and downtime resulting from fires. A fire is very disruptive of production because workers are laid off, or, at best, spend their time cleaning up the mess.

The injuries associated with a fire or during cleanup are very costly and demoralizing. Fires can be prevented by instituting a planned program of inspections and fire drills. Your workers will know that you care when you insist on good housekeeping and that no smoking in certain areas will be enforced.

They will also know they are part of the fire prevention program when they are assigned responsibilities for protection of life and property during fire drills. Additional information regarding fire prevention is available from the National Fire Protection Association.

10 OCCUPATIONAL HEALTH

"The surest road to health, say what they will, is never to suppose we shall be ill."

Churchill

INDUSTRIAL HYGIENE

Occupational health involves a certain professional at the working level known as an industrial hygienist. This professional is concerned with solving occupational health problems with a three-step approach:

- Recognition of the problem
- Evaluation of the environmental factors
- Application of control measures

Many extremely toxic compounds are used daily in industry without being a hazard to your employees. Why? Because precautions are taken to limit actual contact so as not to cause injury or illness.

EXPOSURE TO HEALTH HAZARDS

There are three ways that toxic agents enter the body:

- Inhalation — a great problem — enters the lungs
- Absorption — a minor problem — enters the skin
- Ingestion — a rare problem — enters the stomach

The first step in recognizing potential problem areas in an occupational environment is to become familiar with the particular operation involved. The study involves the toxic materials being used, the way they are used, the number of workers, exposure, and the control measures employed.

The following are airborne contaminants normally found in workplaces:

- *DUST* — solid particles with a wide range of sizes.
- *FUMES* — solid particles formed by the condensation of volatilized solids, usually metals.
- *MIST* — finely divided liquid droplets suspended in air and generated by condensation or atomizing.
- *GAS* — Diffuse formless fluids normally in a gaseous state.
- *SMOKE* — Carbon or soot particles resulting from incomplete combustion of carbonaceous materials.
- *VAPOR* — Gaseous form of substances which are normally in the solid or liquid state.

Regulatory information concerning the above is found in the OSHA Regulations, Title 29, Part 1910.1000. Subpart Z of the Regulation provides a list of allowable limits for toxic and hazardous substances. The list also gives ceiling values and time weighted averages for each hazard.

Chemical Hazards

Chemical hazards arise from excessive airborne concentrations of mists, vapors, gases or solids that are in the form of dust or fumes. These can be inhaled, ingested, or absorbed into the body.

Physical Hazards

Physical hazards include excessive levels of electromagnetic and ionizing radiation, noise, vibration, and extremes of temperatures and pressures.

Biological Hazards

Biological hazards are caused by insects, molds, fungi, and bacterial contamination. Avoidance of these hazards includes monitoring drinking water, sewerage, food handling, sanitation, industrial waste, and personal cleanliness.

Oxygen Deficiency

Deficiency of oxygen in the atmospheres of confined spaces is usually a problem of industry and construction projects. The oxygen content of any tank, vessel and other confined spaces must be checked before entry. Oxygen analyzers are available for this purpose. Entry into confined spaces without a check of oxygen content can cause death. The worker will have no warning, will become unconscious, and will not be able to cry for help.

PROFESSIONAL EVALUATIONS

The person responsible for safety and health in your workplace will look for the causes of work illnesses by checking your daily log of injury/illness, conduct work-site studies, evaluate the effectiveness of exhaust ventilation, and specify what respiratory protection is needed. The professional will also specify hazard controls and solutions for environmental problems that, if left unresolved, would certainly bring OSHA to your door.

The professional will also specify the type of personal protective equipment and clothing required for each hazard. This includes respirators and the type of eye protection necessary to protect the worker when firing a laser beam. Additional tasks will be to conduct a noise survey to determine noise levels and specify control measures. Your safety and health professional will look into human factors and hazard communication.

If your company is small and you are without the service of a trained professional, you should consider contracting the service of an industrial hygiene consultant. Your industrial engineer or plant engineer may serve as your safety and health manager on a collateral duty basis. If that is the case, that person is responsible to see that appropriate information is obtained and action is taken to evaluate and control all health hazards found.

Many occupational diseases cause symptoms similar to non-occupational illnesses. The industrial hygienist can assist a physician by providing information relating to the patient's job. The information enables the physician to correlate the patient's condition with the known health hazards found at the work-site.

Industrial Physicians

The industrial physician is as concerned with your workers' health as the industrial hygienist. The physician will perform pre-placement and periodic examinations of employees working with health hazards. This is very important to you because you need to know about the health of a new employee. It is possible that the new person may have a health problem and not be able to wear a respirator. Also, if you do not examine the new person, you may not be able to prove later on that you did not expose that worker to a health hazard.

For example, how will you be able to prove that you did not contribute to a worker's hearing loss unless you test the person's hearing before hiring them? I am not saying you should not hire a person with a hearing loss, but a baseline examination will help convince the worker to protect his or her hearing from further loss.

The industrial physician works very well with the industrial hygienist. As a matter of fact, they bridge the gap between your workplace and the medical community.

Benefits Achieved

- Good employee health
- Reduced compensation costs
- Lower insurance premiums
- Lower medical expenses
- Increased productivity

YOUR BODY SENSORS

As a primer to help you to detect if something is unsafe or unhealthy, try this when you visit your production or construction sites; try using your built-in body sensors to detect hazards or dangers.

- Ears — Loud noise and discomfort
- Eyes — They will tear, blink and smart
- Mouth — Bad taste and coughing
- Lungs — Will constrict and burn
- Skin — Prickly, perspiration and goose flesh
- Hair — Standing on end
- Nose — Bad smell
- Hands — Clammy, cold or hot
- Tongue — Thickening

The above list will not tell you everything, because many hazards are not detectable except by instrumentation, but they are generally good indicators that you need the expertise of a qualified safety and health professional. Do not take a chance on health hazards. The effects or medical problems may take years before the disease or illness becomes apparent; and when it does, it can bring your company to the wall.

CONSULTATION ASSISTANCE

Free consultation assistance is available to you in establishing a healthful workplace. The consultation service is provided by state government agencies or universities employing health consultants, and they are funded by OSHA. The service is confidential in nature, with the assurance that your company will not be reported to OSHA for purpose of inspection. The service is designed to assist small employers to comply with the health requirements of the OSHA Act. Additional information concerning the consultation assistance can be obtained by requesting OSHA Publication Number 3047.

SUMMARY

The employer is required by law to provide a workplace that is free from recognized health hazards that are likely to cause death, illness, or serious harm to employees. All health hazards are very serious and should not be taken lightly. Untreated, health hazards can haunt you forever. A good example is the asbestos hazard which is bankrupting many businesses. Take a good look at what you do and the health hazards associated with your work processes. It may well be that you are doing everything right. However, if you are not, get help to identify the problem, fix the hazard, and then go on with your business.

11   HAZARD COMMUNICATION

"There is often a sin of omission as well as a sin of commission."

Marcus Aurelius

HAZARD COMMUNICATION STANDARD

The Hazard Communication Standard became law in 1986 and is also known in some states as "The Right to Know". The complete standard is found in the Code of Federal Regulations, Title 29, Part 1910.1200. Hazard communication is changing the way chemicals are handled in businesses and industry.

CHEMICALS IN THE WORKPLACE

Approximately 32 million workers, about 1 in 3 in the nation's workforce, are exposed to one or more chemical hazards. There are an estimated 575,000 existing chemical products and hundreds of new ones being introduced annually. These pose a serious problem for you and your employees.

Chemical exposure may cause or contribute to many serious health effects such as heart ailments, kidney and lung damage, sterility, cancer, burns, and rashes. Some chemicals may also be safety hazards and have the potential to cause fires, explosions, and other serious accidents.

INFORMING THE USER

The purpose of the Hazard Communication Standard is to establish uniform requirements to make sure that the hazards of all chemicals produced, imported, or used within the United States' manufacturing sector are evaluated, and that this hazard information is transmitted to you and your employees.

Chemical manufacturers and importers must convey hazard information to the user employer by means of labels on containers and material safety data sheets (MSDS). In addition, all affected employers are required to have a

Hazard Communication Program to provide the information to their employees by means of container labeling and other forms of warning, MSDS, and training.

This will ensure that you will receive the information you need to inform and train your employees properly, and it will allow you, the employer, to set up and put in place your employee protection programs. It will also provide necessary hazard information to your employees so they can participate in, and support the protective measures that you have instituted in their workplace.

HAZARD EVALUATION

The quality of your Hazard Communication Program is very dependent on the adequacy and accuracy of the initial assessment. The chemical manufacturers and importers of chemicals are required by law to review the available scientific evidence concerning the hazards of the chemicals they produce or import, and to report the information they find to their own employees and to employers who purchase their products.

The chemical manufacturers, importers, and employers are responsible for the quality of the hazard determination they perform. They are also accountable to evaluate each chemical for its potential to cause adverse health effects and its potential to pose physical hazards, such as flammability.

Chemicals which are listed in one of the following sources are considered hazardous in all cases:

- 29 CFR, Part 1910, Subpart Z, Toxic and Hazardous Substances.
- Threshold Limit Values for Chemical Substances and Physical Agents in the Work Environment, American Conference of Governmental Industrial Hygienists.
- Chemicals which have been evaluated and found to be suspect or confirmed carcinogens.

WRITTEN HAZARD COMMUNICATION PROGRAM

All employers must establish a written comprehensive hazard communication program which includes provisions for container labeling, material safety data sheets, and a training program for employees. The written program must include a list of the hazardous chemicals in each work area, the means you will use to inform your employees of the hazards of nonroutine tasks, hazards associated with chemicals in unlabeled pipes, and the way you will inform other contractors working for you at your place of business of the hazards to which their employees may be exposed. You are both responsible for the hazardous chemicals that are brought on site by other contractors.

The written program does not have to be lengthy or complicated, and you may be able to rely on an existing Hazard Communication Program to comply with the above requirements. The written program must be available to your employees, their designated representatives, and to Occupational Safety and Health agencies.

Labels and Warnings

The purpose for labels on hazardous materials containers is clear; it is to provide immediate warning to your employees and a link to more detailed information concerning the chemical which may be found in the Material Safety Data Sheet.

The hazard warning must be specific. The warning must communicate the health hazard of the chemical. For example: "If inhaled, the chemical can cause lung damage."

The chemical manufacturers, importers, and distributors must label, tag, or mark all containers with the identity, appropriate hazard warning, and the name and address of the manufacturer or other responsible party.

You are not required to label portable containers into which hazardous chemicals are transferred from labeled containers, but for the sake of consistency, it may be a good policy to do so. Years ago, a worker transferred toxic substances into a soft drink bottle, which was then swallowed by another worker. You can guess the rest of the story. The law does not require labeling all piping systems, but again, it would be prudent to do so because many State Right-to-Know public laws require that piping systems be labeled.

Material Safety Data Sheets (MSDS)

The chemical manufacturers and the importers of chemicals must develop MSDS for each hazardous chemical they produce or import. As the user of chemicals, you are responsible for obtaining a MSDS for each hazardous chemical or substance used in your workplace.

Beyond identity information, chemical manufacturers and importers must also provide the following information for each chemical:

- Physical and chemical characteristics
- Known acute and chronic health effects
- Related health information
- Exposure limits
- Carcinogenic potential
- Precautionary measures
- Emergency and first aid procedures
- Fire and explosion data
- Reactivity data

- Spill or leak procedures
- Permissible exposure limits
- Hazardous ingredients
- Special precautions
- Special protection

Copies of the MSDS must be readily available and accessible to your employees. They must also be available close to the worker during each shift.

EMPLOYEE INFORMATION AND TRAINING

The employer must establish a training and information program for employees exposed to hazardous chemicals in their work areas at the time of initial job assignment, and whenever a new chemical hazard is introduced into their work area. The discussion topics must include, at least:

- The existence and requirements of the Hazard Communication Standard.
- The components of the Hazard Communication Program.
- Operations in the employees work area where hazardous chemicals are present.
- Where the employer will be keeping the written procedures, programs, lists of hazardous chemicals, and the associated Material Safety Data Sheets.

The training plan must consist of:

- How the Hazard Communication Program is implemented in the workplace.
- How to read and interpret information on labels and MSDS.
- How employees can obtain and use the available hazard information.
- The hazard of the chemicals in the work area.
- Measures the employees can take to protect themselves from the hazards.
- Specific procedures put into effect by the employer to provide protection, such as work practices.
- The proper use of personal protective equipment.
- Methods and observations, such as visual appearance or smell that workers can use to detect the hazard.

TRADE SECRETS

There are provisions in the Standard for "trade secrets". The Standard strikes a balance between the need to protect your exposed employees and the employer's need to maintain the confidentiality of a "bona fide trade secret", (see OSHA, Title 29, Part 1910.1200). The chemical manufacturer, importer or employer must immediately disclose the specific chemical identity of a hazardous chemical to a treating physician or nurse when the information is needed for proper emergency or first aid treatment. As soon as circumstances permit, the chemical manufacturer, importer, or the end user employer may obtain a written statement of need and confidentiality statement.

Under the contingency described here, the treating physician or nurse has the ultimate responsibility for determining that a medical emergency exists. At the time of emergency, the professional judgment of the physician or nurse regarding the situation must form the basis for triggering the immediate disclosure requirements.

In nonemergency situations, chemical manufacturers, importers, or end use employer must disclose the withheld specific chemical identity to health professionals providing medical or other occupational health services to exposed employees if certain conditions are met. In this context "health professionals" include: physicians, industrial hygienists, toxicologists, or epidemiologists.

WRITTEN REQUEST

The request for information must be in writing and must describe with reasonable detail, the medical or occupational health need for the information. The request of the health professional will be considered if the information will be used for one or more of the following activities:

- To assess the hazards of the chemical to which employees will be exposed.
- To conduct or assess sampling of the workplace atmosphere to determine employee exposure levels.
- To conduct pre-assignment or periodic medical surveillance of exposed employees.
- To provide medical treatment to exposed employees.
- To select or assess appropriate personal protective equipment for exposed employees.
- To design or assess engineering controls or other protective measures for exposed employees.
- To conduct studies to determine the health effects of exposure.

The health professional must also specify why alternative information is insufficient, must explain in detail why disclosure of the specific chemical identity is essential, and including the procedure to be used to protect the confidentiality of the information. It must also include an agreement not to use the information for any purpose other than the health need stated.

SUMMARY

The Hazard Communication Standard not only safeguards your employees, but it protects your business from disasters. In the past, many people bought chemicals and compounds for what they could do, and not for what they

were. Training your employees to work with chemical hazards is very important to the success of your business. Do not make the training complicated to
the point that your workers are overwhelmed, but be specific. The training
should include how to detect a chemical release, the physical and health
hazards of each chemical used, the protective measures to limit exposure, the
protective equipment required, and the details of the Hazard Communication
Standard, including labels, warning, and the Material Safety Data Sheets.

If you have not yet prepared a written program, it is never too late to do
so. Remember, it is the law. You should begin by making a master listing of
all the chemicals in use at your workplace. You should then obtain the MSDS
from your supplier for each chemical substance used. If during the audit of your
workplace, you note deficiencies in the MSDS, contact your vendor of chemicals to request an updated MSDS.

The Hazard Communication Rule applies to those in manufacturing, workers
in any industry ranging from construction and transportation through the
wholesale and retail trades. The rule also applies to the service industries when
and if they are potentially exposed to hazardous substances.

12 RESPIRATORY PROTECTION

"For life is not to live, but to be well."

Martial

A REQUIREMENT OF LAW

The regulations regarding the control of occupational diseases caused by breathing air contaminated with harmful dusts, fogs, fumes, mists, gases, smokes, sprays, or vapors can be found in the OSHA Regulations, 29 CFR, Part 1910, and its subparts.

PROTECTION OF EMPLOYEES

Because there are hundreds of operations requiring respiratory protection, I will not attempt to list all operations and the type of respirator required for each one. Instead I will tell you when protection is required and leave the selection to you, your consultant, or to your local safety equipment supplier.

The Occupational Safety and Health Act requires that employees shall be protected from breathing air containing hazardous concentrations of dust, fumes, mists, gases or vapors. The primary means to control or remove these hazardous concentrations is by engineering methods. The way to do this is to use exhaust ventilation. When this is not feasible, employees must be protected by the use of respirators.

In either case, your workplace should be evaluated by a professional, one who is qualified in respiratory protection and industrial ventilation. The consultant will specify the type of respirator required for each hazard and/or the type of ventilation required, and when you have done this, you will be keeping your employees from breathing hazardous air.

SELECTING A RESPIRATOR

Respiratory protection devices vary in design and their usage. Selection depends on the toxic substance that is in the breathing air. The selection of a respirator should be made according to the guidelines in the American National Standards Practice for Respiratory Protection (ANSI Z288.2-1969).

RESPIRATOR ISSUE AND CARE

Respirators should be issued from a central area where they will be cleaned and maintained. Usually, the person in charge of your tool room can be given the task to dispense the respirators. This person should not issue respirators to employees with beards or long sideburns, because the wearer of the respirator will not be able to achieve a proper face seal.

Respirators must not be left in personal lockers. They should be returned to the issue room regularly for repair and sanitizing. Replacement of other than disposable parts and any repair should be done only by a person trained to insure that the equipment is functionally sound.

TRAINING THE EMPLOYEE

The user of the respirator must be instructed and trained in the proper use and limitations of a respirator. Workers should not be assigned to tasks requiring use of respirators unless it has been determined that they are physically able to perform the work and use the equipment. Your company physician will determine which health and physical conditions are pertinent.

Every respirator wearer needs to receive fitting instructions, including demonstrations and practice in how the respirator should be worn, how to adjust it, and how to determine if it fits properly. Respirators must be NIOSH-MESA approved.

As the employer, you are responsible to have a written standard operating procedure governing the selection and use of respirators. You are also responsible for performing regular inspections and evaluating the respirator program to determine its effectiveness.

Unless you make a determined effort to remove air-borne contaminants by the use of ventilation, you will be required by law to maintain an effective respiratory protection program for your employees. The companies that have approved ventilation systems to remove the dust, fumes, mist, gas, smoke, etc., are able to minimize the use of respiratory protection equipment.

TYPES OF RESPIRATORS

The following are the respirator types available from a safety equipment supplier. You may be able to select the type necessary for your particular kind of hazard. However, it is a good idea to check with your consultant before you decide to purchase something that will not work and could endanger your employees.

- Self-contained breathing apparatus (rescue)
- Air supplied respirator
- Powered air purifying respirator
- Fume, dust, and mist filter respirators
- Dust and mist filter respirators
- Specific chemical hazard cartridge respirator
- Combination mechanical/chemical filter respirator
- Gas masks with appropriate canister
- Emergency escape breathing devices
- Combination supplied air and escape respirator
- Air-fed hood-type respirator

WORD OF CAUTION

The above list of respirators is by no means a complete list of available respirators, nor is it my intent that you should use the above to select a respirator. Also, you should not allow your supervisors to determine the protection required unless they are qualified. Why? Because a dust filter provides no protection against chemical vapors. An organic vapor cartridge provides very little protection against hazardous dusts.

Combinations of hazardous substances in different physical states require an air-fed respirator. There are many possibilities in various workplaces where more than one air contaminant in different physical states may be found. While some respiratory hazards are recognizable, there are others that are invisible and impossible to detect by odor, taste, or irritation. Do not take a chance that could be very costly to you and your employees.

MEDICAL RESTRICTIONS

You should not allow your employees to wear a respirator for the following reasons:

- Respiratory problems such as asthma, severe allergies, or emphysema
- Circulatory problems such as high blood pressure or heart disorders
- Psychological problems such as claustrophobia
- Facial problems such as scars or excessive facial hair

Contact lenses should *not* be worn when wearing a respirator because they might be dislodged through pressure changes or they could cause irritation if a contaminant gets into the eyes. The temple bars of corrective eyeglasses will interfere with the seal of some respirators, however, the supplier of respirators can provide you with a spectacle kit, or you can allow your employees who need to wear corrective eye wear to shorten the temple bars and tape them to their heads. Your employees must receive an annual physical examination prior to wearing a respirator.

MINIMAL REQUIREMENTS (OSHA)

- Establish a written operating procedure for respirator selection and use.
- Respirator selection must be based on the hazards to which the worker is exposed.
- The user must be instructed and trained to use the respirator properly and recognize its limitations.
- The respirator should be assigned to an individual for his or her exclusive use, if possible.
- Clean and disinfect respirators regularly, after each day's use, or if used by more than one worker.
- Store respirators in a convenient, clean, and sanitary place.
- Inspect routinely used respirators at least once a month and after each use.
- Maintain surveillance on work area conditions and degree of employee exposure or stress.
- Regularly inspect and evaluate the continuing effectiveness of the program.
- Allow only those persons who are physically able to perform work requiring the use of a respirator.
- The respirator user's medical status should be reviewed periodically.

LOCAL EXHAUST VENTILATION

A local exhaust ventilation system is used to carry off air contaminants by trapping the contaminants near its source. For example, welding may produce fumes and gases hazardous to your employees' health. The exhaust ventilation system consists of a hood close to the work to be performed, ducts for carrying the contaminated air to a central point, an air cleaning device to clean the air before it is discharged, and a fan and motor to create the required air flow throughout the system.

It is very important that you consult with an expert regarding placement, size of ducts, and capacity of fan, motor and filter. Do not confuse local exhaust ventilation with general ventilation, such as that used for heating, air conditioning, make-up air, and other comfort ventilation.

You should never attempt to hook general ventilation with local exhaust ventilation. The ventilation requirements may be found in the American National Standards Institute Instruction, ANSI Z249.1, and the OSHA Regulations, 29 CFR, Part 1910.

SUMMARY

Removing the unwanted air-borne contaminants from your work-sites is not only a requirement of OSHA, but it makes good sense, and it will increase the productivity of your employees. There may be situations where this is not feasible; in that case, you are required by law to provide respiratory protection.

It is entirely possible, due to the nature of your business, that you are required to have both respiratory protection and local exhaust ventilation. However, you will only know for sure what is required when you hire a consultant with experience in ventilation and respiratory protection. Take a good look at your work-sites, and if what you see is hazy, smoky, and has a bad smell, you either do not have an exhaust ventilation system, or if you do, it is not functioning properly.

13   PROTECTIVE EQUIPMENT

"Confident because of our caution."

Epictetus

PERSONAL PROTECTIVE EQUIPMENT

Personal protective equipment (PPE) includes all clothing and accessories designed to create a barrier against workplace hazards. Some examples are as follows:

- Head protection — hard hats
- Eye protection — safety glasses
- Face protection — face shields
- Ear protection — ear plugs
- Body protection — special suits
- Hand protection — gloves
- Foot protection — safety shoes

Many years ago, I remember an old safety specialist who carried a glass eye in his pocket. He told a group of young workers a few words that I have never forgotten, "You can walk with a wooden leg, you can hear with an aid, you can chew with false teeth, but you'll never see through a glass eye."

PROTECTING THE EMPLOYEE

The basic element of any PPE management program should be an in-depth evaluation of the equipment needed to protect your employees from the hazards found in your workplace. Do not go out and buy protective equipment for the sake of having some equipment. Instead, do an evaluation and then consult with a safety professional. The professional will save you money by recommending equipment that will work.

Your management staff, dedicated to the safety of your employees, should use the above evaluation to set a standard operating procedure for your employees. In other words, they should ensure that your employees are trained to use the equipment to protect themselves against the hazards. Using personal protective equipment requires hazard awareness and training on the part of the user; but a word of caution, your employees must be made aware that the equipment does not *eliminate* the hazards.

Selection of the proper piece of personal protective equipment for the job is very important. Your workers must understand the equipment's purpose and limitations. The equipment must not be altered or removed, even though an employee may find it uncomfortable. Sometimes equipment may be uncomfortable simply because it does not fit properly.

Head Protection

Prevention of head injuries is an important factor in your safety program, because a survey by the Bureau of Labor Statistics of Accidents and Injuries noted that most workers who suffered impact injuries to the head were not wearing head protection at the time of injury. An important point in the survey shows that most employers did not require their workers to wear head protection.

It was also found that the vast majority who wore them all or most of the time, felt that hard hats were practical in their jobs. In almost half of the accidents involving head injuries, the worker knew of no action taken by the employers to prevent such injuries from occurring again. Because it is difficult to anticipate when and where head injuries will occur, head protection should be worn whenever there is even a remote chance that a head injury could happen.

Head injuries are caused by falling or flying objects, or by bumping the head against a fixed object. Hard hats must do two things — resist penetration and absorb the shock of a blow. This is accomplished by making the shell of the hat of a material hard enough to resist the blow, and by utilizing a shock absorbing lining composed of a headband and crown straps to keep the shell away from the worker's skull.

Eye and Face Protection

Eye and face protection is required where there is a reasonable probability of preventable injury to your workers. This requirement must also apply to your supervisors, management personnel, and visitors while they are in areas that contain hazards. The Bureau of Labor Statistics study found that 60% of workers who suffered eye injuries were not wearing eye protection equipment. When asked why they were not wearing protective equipment at the time of the accident, workers indicated that eye and face protection equipment was not normally used or practiced in their type of work, or it was not required for the type of work performed at the time of the accident.

You must provide suitable eye protectors where machines or operations present the hazard of flying objects, glare, liquids, radiation, or a combination of these hazards. Consideration should be given to the type and degree of hazard, and the kind of protection required should be selected on that basis.

Persons using corrective spectacles and those who are required by OSHA to wear eye protection must wear face shields, goggles, or spectacles of one of the following types:

- Spectacles with protective lenses providing optical correction
- Goggles worn over corrective spectacles
- Goggles that incorporate corrective lenses mounted behind the protective lenses

Many hard hats and nonrigid helmets are designed with face and eye protective equipment.

Hearing Protection

Exposure to high noise levels can cause hearing loss or impairment to your employees. It can also create physical and psychological stress. There is no cure for noise-induced hearing loss. So, prevention of excessive noise exposure is the only way to avoid hearing damage. Specifically designed protection is required, depending on the type of noise encountered.

Pre-formed or molded earplugs should be individually fitted by a professional. Waxed cotton, foam, or fiberglass wool earplugs are self-forming. When properly inserted, they work as well as most molded earplugs. Some earplugs are disposable — to be used one time and then thrown away. The nondisposable type should be cleaned after each use for proper protection. Plain cotton wads are ineffective as protection against hazardous noise.

Earmuffs need to make a perfect seal around the ear to be effective. For extremely noisy situations, you should require your employees to wear earplugs in addition to earmuffs. This is known as double protection and when used together, earplugs and earmuffs change the nature of sounds; all sounds are reduced, including one's own voice, but other voices or warning signals are easier to hear. Of course, the best defense against noise is to sound-dampen the equipment, or isolate the noise from other workers with sound isolation barriers.

Body Protection

There are many hazards that can harm the body, and are caused by high heat, hot liquid metal, cuts, acids and radiation. In order to protect your workers from those hazards, a variety of protective clothing is available in the form of vests, jackets, aprons, coveralls, and full-body suits. Some protective clothings are fire resistant and comfortable to wear, since they adapt well to changing temperatures.

Duck, a closely woven cotton fabric, is good for light-duty protective clothing. It can protect against cuts and bruises on jobs where employees handle heavy, sharp, or rough materials. Heat-reflecting clothing such as leather, is often used to guard against dry heat and flame. Rubber and rubberized fabrics, neoprene, and plastics give protection against some acids and chemicals.

Disposable suits of paper-like material are particularly important for protection from dusty materials or materials that can splash. If the substance is extremely toxic, a completely enclosed suit may be necessary. The clothing should be inspected to assure proper fit and function for continued protection.

Arm and Hand Protection

Examples of injuries to your employees arms and hands are burns, cuts, electrical shocks, amputation, and absorption of chemicals. There are a wide assortment of gloves, hand pads, sleeves, and wristlets for protection from various hazardous situations.

The protective device should be selected to fit the job. For example, some gloves are designed to protect against specific chemical hazards. Employees may need to use gloves which have been tested and provide insulation from burns and cuts, such as wire mesh, leather, or canvas. Your employees should become acquainted with the limitations of the clothing used.

Certain occupations call for special protection. For example, electricians need special protection from electrical shocks and burns. Rubber is considered the best material for insulating gloves and sleeves for electrical workers.

Foot and Leg Protection

According to the Bureau of Labor Statistics, most of the workers in selected occupations who suffered impact injuries to the feet were not wearing protective footwear. Furthermore, most of the employers did not require their workers to wear safety shoes. The typical foot injury was caused by objects falling less than 4 feet and the average weight of the object was about 65 lb. Again, most workers were injured while performing their normal job activities at their work-sites.

For protection of feet and legs from falling or rolling objects, sharp objects, molten metal, hot surfaces, and wet slippery surfaces, your workers should use appropriate foot guards, safety shoes, or boots and leggings. Safety shoes should be sturdy and have impact-resistant toes. Additional protection such as metatarsal guards may be found in some types of footwear. Leggings protect the lower leg and feet from molten metal or welding sparks. Safety snaps permit their rapid removal.

Other Protection

There are many other types of protective equipment available for your employees to wear, but they are too numerous to mention. I will, however, list a few to give you an idea and some extra thoughts regarding the protection of your workers.

- Coast Guard-approved life jackets or vests when working over or near water
- Hair nets when working around machinery
- Lifelines, safety belts, body harnesses, lanyards and safety nets when working aloft
- Respirators when working in hazardous atmospheres
- Barrier creams for hands when working with defatting substances

SUMMARY

Personal protective equipment can be very effective in reducing injuries when the protective equipment is selected based on its intended use, when employees are trained in its use, and when that equipment is properly maintained. Remember, you are required by law to provide protection for your employees and to ensure that the equipment is worn by your workers.

It will not help your business if your workers do not bother to use the safety equipment and your injuries continue to increase. When this happens, you are paying twice, once when you buy the equipment to prevent injuries, and again when you pay for the injuries that could have been prevented by the enforcement of wearing personal protective equipment. When you provide protective equipment but do not enforce its use, way down the pike you will be paying for hearing loss, loss of sight, and many other injuries and long-term illnesses.

14 SYSTEM SAFETY

"There must be a better way — find it."

Thomas Edison

A NEW APPROACH

The better way is system safety. It is not new. As a matter of fact, the Department of Defense, in 1969, developed a military standard known as Mil-STD-882 and called "System Safety Program for Systems and Associated Sub-Systems and Equipment". The motivating force to produce this standard started when the Department of Defense reported heavy losses to aircrafts, submarines, missiles, and ground vehicles.

A TOTAL SAFETY PROGRAM

System safety is a conscientious effort to avoid accidents by eliminating or reducing hazards associated with all work systems. It is a known fact that most accidents result from lack of planning, starting at the conceptual phase, right on through to the completion or operational phase of a project.

Remember this, accidents do not just happen by themselves, but instead are caused by unplanned and unforeseeable events. Another recognized fact is, the most effective way to avoid accidents that lead to injury, death, and damage to equipment is to anticipate situations prone to change during the many phases of a project or task. For example, operating a pump to pressure test an unfired pressure vessel without installing an overpressurizing device could expose your workers to great danger, not to mention equipment damage; or using chemicals without reviewing the Material Data Sheets (MSDS), is chance taking, and a violation of the Hazard Communication Law.

ANTICIPATING ACCIDENTS

The most effective means to avoid accidents is by eliminating or reducing hazards and dangers during system planning or development. Simply said, all those associated with a project or task must reform their input. You can do this by studying each step or phase in order to anticipate situations prone to accidents. The following phases will apply to any task or project:

- Conceptual phase
- Design phase
- Material acquisition phase
- Fabrication phase
- Assembly phase
- Testing phase
- Operational phase
- Maintenance phase

Oftentimes, hazards are not detected until an accident occurs — a costly oversight. Trying to anticipate failures or detecting hazards prior to the operational phase is not always foolproof, but the exercise is considered essential for the preservation of human and material resources.

SAFETY TRAINING

It is very important that you review the total concept of system safety, which involves interaction between people, machines, products, and the environment. During the system safety analysis of a task or project, careful consideration must be made to ensure the integration of people into the system. You should also recognize the performance imperfection of your employees. Imperfect performance, a concern of all managers, becomes the connecting link between accident prevention and getting the job complete on time, without an accident or injury.

The analysis must ensure that your people have received training in the use of tools, machines, hazardous materials and the environment in which they are required to work. The point is raised here to alert preparers of work instructions to include safety and health input, because failure to do so involves correcting each employee for system failure and does little to improve problems generated by planning and training oversight.

THE HUMAN ASPECT

During each safety review of each phase, it is vital to consider the human aspect such as, was the system designed to fit the people? All too often, the tendency is to design the system and then later on, the people are back-fitted as an afterthought. This practice is not only ineffective, but sometimes very costly. Emphasis on human factors found in Chapter 8 can be very helpful during the conceptual phase.

The science of human factors attempts to design equipment that can be operated easily and rapidly with a minimum of undue strain. Therefore, in order to obtain maximum reliability, the best capabilities of people and machines must be integrated into the system during each phase. The process should anticipate misuse, the consequences of normal and abnormal wear, and servicing/repair requirements. There are times when equipment that has been recognized as being hazardous will be accepted as part of the risk associated with the task. When this happens, the only approach to safeguard your employees is to isolate the people from the hazard, or the hazard from the people.

FAILURE ANALYSIS

System safety analysis must be performed in order to identify new hazards as a result of engineering changes and upgrading or alterations. Some changes may be the result of failure analysis performed during accident investigations. Also, do not overlook a safety review of operating and maintenance procedures when they are revised because of the above changes. The review will ensure that safety is not degraded as a result of modifications. In the past, accident investigations uncovered the following failures:

- Failure to keep up with safer technology
- Failure to warn of built-in hazards
- Failure to utilize safe materials
- Failure to install safety devices
- Failure to anticipate wear
- Failure to anticipate environmental changes
- Failure to anticipate greater use
- Failure to protect employees from harm

System safety is a composite of safety inspection, accident investigation, human factors, job hazard analysis, safety training, and personal protective equipment.

LEARNING FROM MISTAKES

An accident or injury is a tough instructor, the test is first and the lesson is last. If you do not learn from your mistakes, you are bound to suffer many repeat accidents. Take a hard look at your accidents and injuries. They will tell you just how your system is working. It makes no difference if you are running a production line, erecting a building, or building a ship; a system is a system.

It may be impossible to reduce the risk of accidents to zero, but you should minimize the risk to a manageable level. A system is safe if its risk is judged acceptable from the point of preventing accidents and injuries. Driving your automobile down the highway is a risk, but driving imprudently or at high speeds is chance taking. Risk is a control while chance is out-of-control.

SYSTEM SAFETY CHECK LIST

Probably the best way to start a system safety analysis would be to construct a check list of all support systems used for each phase of a project or task. An example of the system employed could be as follows:

- Mechanical
- Structural
- Electrical
- Electronic
- Chemical
- Environment
- Material handling
- Motorized equipment
- Hoisting
- Piping systems
- Pressurized vessels

I will not attempt to list all the possible systems that exist, such as lasers, nuclear, and the many others that are in use or those in the conceptual phase. However, Figure 1 for electrical systems will give you an example of a checklist or analysis of the thought process you should use when working with electrical energy. Again, I am sure you can add other items to the electrical system safety analysis.

You may want to use Figure 1 as a format when conducting a job hazard analysis in Chapter 6. Another good reminder, is that each chapter in this book supports each other with common goals to prevent accidents and injuries. Some people will say that a check list is a waste of time. I disagree!! We all use check lists when we plan even the smallest task. It also helps us to estimate what it will take and what it will cost for each project. So why not take the time to avoid costly oversights that could result in a disaster?

EVOLVING TECHNOLOGY

As technology advances, machines become increasingly complex and efficient, and better able to replace humans in production processes. However, as technology advances, there remain two fundamental problems to be dealt with. The first is the obvious fact that, even though machines are improving constantly, machines are still machines and parts eventually wear out.

Potential Hazard	Potential Accident	Adverse Effects	Checkpoints	Support Action	Protective Equipment
Working on energized electrical circuits	Electrical shock Electrocution Electrical fires Explosion	Burns Serious injury Fatality Damage to equipment	Proper grounding Check circuits Install signs, tags and barriers Availability of fire extinguisher Insulate metal tools Use only approved equipment and tools Provide lighting Check walking/ working surfaces Check lockout devices Check overcurrent protection devices Check ground fault circuit interrupters Availability of rubber insulation Approved electrical equipment Inspect electrical component instal- lation Check guards	CPR and first- aid trained Trained on hazards of electricity Tag-out/lock- out control Other emer- gency pro- cedures Communication Use approved equipment Remove metal objects from body	Utilize rubber protective devices Wear eye pro- tection Wear hand pro- tection Utilize safety lines Use leather pro- tection Wear head pro- tection Wear rubber footwear on wet surfaces

Figure 1

As far as the future of system safety, it is enlightening to see that many industries fulfill two functions daily. Their regular function to make a product and the safety aspect of that product.

SUMMARY

System safety is a thought process during each phase of a project. It is also a last look at a project or task to see if everything fits together. It is asking if quality is assured, because without quality, safety is unsure.

System safety analysis is needed to find out if people are trained to operate the equipment, if human limitations and comfort were considered, and what type of protective equipment was specified.

The maintenance, alterations, and engineering changes are all factors incorporated into the system safety review. In fact, it is an exercise in perform- ing a fail-safe system review that will result in quality and a safe product. As a last reminder, fault the system if you will, and not the people in it, for lasting results.

15   EMERGENCY RESPONSE

"Thou shalt show the people the way wherein they must go."

Exodus

PLANNING FOR EMERGENCIES

Regardless of how effective you think your safety and health program is, accidents still occur in spite of efforts to prevent them. Therefore, proper planning for emergencies is necessary to minimize employee injury and property damage. This chapter details the basic steps needed to prepare you to handle emergencies in your workplace.

PURPOSE OF RESPONSE

Emergencies include, but are not limited to, accidental releases of toxic gases, chemical spills, fires, explosions, and personal injury. While many companies already have programs in effect, for these companies this chapter can reinforce existing programs.

The effectiveness of response during emergencies depends on the amount of planning and training performed. If you believe that your safety and health program is very important, you will also see that emergency response is equally important.

Some emergency action plans are required by certain OSHA standards. When they are required by OSHA, they must be in writing; except for firms with 10 or fewer employees, then the plan may be communicated orally to your employees.

EMERGENCY ACTION PLAN

The emergency response or emergency action plan must include, as a minimum, the following elements:

- Emergency escape procedures and emergency escape route assignments
- Procedures to be followed by employees who remain to perform (or shut down) critical plant operations before they evacuate
- Procedures to account for all employees after emergency evacuation has been completed
- Rescue and medical duties for those employees who are to perform them
- The preferred means for reporting fires and other emergencies
- Names or regular job titles of persons or departments to be contacted for further information or explanation of duties under the plan

The emergency action plan should address all potential emergencies which can be expected in your workplace. Therefore, it will be necessary to perform a hazard audit to determine potentially toxic materials and unsafe conditions. See Chapter 4, Safety Inspections and Chapter 11, Hazard Communication. These two chapters will give you all the information you will need to conduct an audit.

You should list in detail the procedures to be taken by those employees who must remain behind to care for essential plant operations until their evacuation becomes absolutely necessary. This may include monitoring plant power, water, and other services that cannot be shut down for every emergency alarm.

For emergency evacuation, the use of floor plans or workplace maps which clearly show the emergency escape routes and safe or refuge areas should be included in the plan. Your employees must be told what actions they are to take in the emergency situations.

WHO TAKES CHARGE

A chain of command should be established to minimize confusion so that employees will have no doubt who has authority for making decisions. Responsible individuals should be selected to coordinate the work of the emergency response team. You may want to assign your plant manager, or some other manager, to be the "coordinator in charge" of the emergency response team. The duties of the coordinator should include:

- Assessing the situation and determining whether an emergency exists.
- Directing all efforts, including evacuating personnel and minimizing property loss.
- Ensuring that outside emergency services such as medical aid and local fire departments are called in when necessary.
- Directing the shutdown of plan operations when necessary.

COMMUNICATIONS

During a major emergency involving a fire or explosion, it may be necessary to evacuate offices, in addition to manufacturing areas. Your normal services, such as electricity, water, and telephones, may be nonexistent. Under these conditions, an alternate area may be necessary where employees can report or which can act as a focal point for incoming and outgoing calls. The coordinator can also use this alternate area to direct his emergency efforts.

Beside emergency means of communication, a method of communication is also needed to alert employees of the evacuation. Alarms should be audible to all the people in the plant. The warning plan should be in writing and you should make sure that each employee knows what it means and what action is to be taken. You will also need to account for all your employees during emergencies.

EMERGENCY RESPONSE TEAMS

Emergency response teams are the first line of defense in emergencies. Before assigning people to these teams, you should assure that your employees are physically capable of performing the duties which may be assigned to them. Depending on the size of your plant, there may be one or several teams trained in the following areas:

1. Use of various types of fire extinguishers (see Chapter 9, Fire Prevention)
2. First aid, including cardiopulmonary resuscitation (CPR), (see Chapter 21, First Aid and Medical Personnel)
3. Shut-down procedures
4. Evacuation procedures
5. Chemical spill control procedures
6. Use of self-contained breathing apparatus
7. Search and emergency rescue procedures

Emergency response teams should be trained in the types of possible emergencies and the emergency actions to be performed. They should be informed about special hazards, such as storage and use of flammable materials, toxic chemicals, radioactive sources, and water-reactive substances, to which they may be exposed during fire and other emergencies (see Chapter 11, Hazard Communication).

TRAINING EMPLOYEES

Training is important to the effectiveness of an emergency plan. In addition to the specialized training for emergency response team members, all employees should be trained in:

1. Evacuation plans
2. Alarm systems
3. Reporting procedures for personnel
4. Shut down procedures
5. Types of potential emergencies

Drills should be held for all personnel, at random intervals, annually at the very least. The emergency plan should be reviewed periodically and updated.

PERSONAL PROTECTIVE EQUIPMENT

Effective personal protective equipment is essential for any person who may be exposed to potentially hazardous substances. Some examples of hazards encountered may be: chemical splashes, falling or flying objects, unknown atmospheres, toxic gases, vapors, mist, electrical fires, and flooding (see Chapter 13, Protective Equipment).

You will most likely need professional consultation in providing adequate respiratory protection, especially for breathing escape masks, and you will need to determine if the employees are physically able to use the respirators, and that they are trained in their use.

MEDICAL ASSISTANCE

In a major emergency, time is a critical factor in minimizing injuries. Most small businesses do not have a formal medical program. However, they *are* required to have first aid services (see Chapter 21, First Aid and Medical Personnel).

SECURITY

During an emergency, it is often necessary to secure the area to prevent unauthorized access and to protect vital records and equipment. An off-limit area must be established by cordoning off the area with ropes and signs. It may be necessary to notify local law enforcement officials or to employ private security to secure the area and prevent the entry of unauthorized personnel.

Certain records may also need to be protected, such as essential accounting files, legal documents, and lists of employees' relatives to be notified in case of emergencies.

Credit for some of the information here is given to the U.S. Department of Labor and OSHA. More information regarding how to prepare for workplace emergencies may be found in OSHA booklet number 3088.

SUMMARY

Planning for emergencies makes good sense. Successful business people have always planned how they should run their companies. It stands to reason that disasters, no matter how small, can also ruin your company. When they occur, small emergencies prepare you to avoid catastrophic accidents.

There are tangible benefits besides complying with OSHA when you plan for emergencies. It not only helps you to prevent disasters, but it will support your safety and fire prevention programs when your employees see that you are serious when it comes to their safety, especially during fire drills and evacuations.

Emergency response also applies to hazardous waste operations which can be found in OSHA 29CFR, Part 1910.120.

16   SAFETY COMMITTEES

"Solutions to problems are like keys in locks; they don't work if they don't fit."

Robert Mager

ACTIVE PARTICIPATION

In the past, safety committees have sometimes been referred to as "management's secret weapon". Why safety committees were called a secret weapon is beyond my recollection. However, Douglas McGregor, in his book titled, *The Human Side of Enterprise* said, "Participation is one of the most misunderstood ideas that has emerged from the field of human relations. It is praised by some, condemned by others and used with considerable success by still others."

There is no doubt that participation has the unique characteristic of giving workers a chance to be part of the action, as well as providing an opportunity for them to contribute to the final solution. On both of these counts, participation is one of the more useful leadership techniques, but it should be added that it must be real participation and not a sham.

If your supervisors are going to invite participation, they must be ready to take it seriously and take action accordingly. There is no doubt that in many organizations the safety committee has been a worthwhile part of the accident prevention program. However, supervisors should not be content to accept the value of safety committees just because of past success. The meetings have to be meaningful; it is a neverending process. There is always room for improvements in any safety situation.

Meetings to solve safety problems can be accomplished by using other forums, such as quality circles or any other problem-solving groups. Safety is everybody's business, therefore, any problem-solving group can take on safety problems, whether or not it is called a safety committee.

SAFETY CAN BE CONTAGIOUS

Many, many years ago, a man with 12 disciples started a religious program that has changed the way we think. One might say, what does this have to do with safety? Well, an active safety committee operates much the same way in spreading or sowing good safety attitudes.

Now, if I were asked, "What's the best way to spread the philosophy of safety?" I would have to say, "Safety committees". Webster's Dictionary describes a disciple as "one who accepts and assists in spreading the teachings of another." That is exactly what an active safety committee does, and much more:

- They create and maintain an active interest in safety.
- They serve as communicators between management and the employees.
- They help set safety policies in motion.
- They arouse and maintain interest in safety.
- They motivate and convince employees to cooperate in preventing injuries.

THE VALUE OF A SAFETY COMMITTEE

It is OSHA's opinion that a safety and health committee, when properly run, can be used to fight more than one battle for management. Once you get employees thinking about the workplace from management's point of view, and if you listen when they talk to you, you will find that they will come up with some excellent suggestions, ranging from safety and health improvements to time saving, to process improvement.

The cost of plant insurance premiums and workers' compensation insurance may be lowered as a benefit of committee work. Many employers have reaped substantial savings after energetic committees brought about a drastic reduction in accident and injuries. Genuine support for a working safety committee also provides an opportunity to demonstrate good faith when the OSHA compliance office comes to your place.

What kind of committee do you need? How many members should be on it? These are among the first questions that come up when management decides to tackle plant hazards through an active safety committee. The answer most frequently given by experts: let the size of the company or plant and its hazard potential dictate the type and size of the committee.

HOW TO ORGANIZE THE COMMITTEE

To command respect, the committee leader or chairman should be a person whose authority exceeds the authority of each member of the group. This gives a fair guarantee of: (1) effective, controlled action to follow committee findings;

and (2) access to the next higher level of management via the committee chairman. Following this general rule, the network of safety and health committees can be set.

Some might say, why do we need a safety committee when we have safety and health professionals doing the same thing? In the first place, your safety staff is busy doing their everyday work. The work of the committee looks at ways to improve the safety and health climate of the organization and to involve the employees in the process. A cardinal rule of good committee work is to keep the group small so that every member can participate actively.

If you have a safety manager, make him or her an advisor to the committee. Also, invite medical people to the meeting. If you are experiencing a rash of fires, invite a fire chief. Many companies have a system of sending copies of minutes to key people throughout the organization in order to keep them posted on the efforts and achievements of the committee. Remember, the most important ingredient is management's commitment. Nothing is more worthless than a showpiece safety and health committee.

FUNCTIONS OF SAFETY COMMITTEE

A safety committee can be very helpful, especially when they are allowed to do the following:

- Discuss safety policies and recommend adoption.
- Help management to correct unsafe conditions and unsafe acts.
- Help to put safety policy and rules into practice.
- Create more disciples of safety.
- Help evaluate safety suggestions.
- Arouse interest in safety.
- Assist supervisors and instructors.

SAFETY COMMITTEE PROCEDURE

When a committee is formed, a procedure should be written describing its function and activity.

- Extent of committee authority
- Scope of committee activity
- Order of business
- Record-keeping
- Attendance

The committee should keep records of activity in writing. However, do not build it up, keep it simple and stick to the facts.

COMMITTEE MEETINGS

The frequency of meetings depends on the tasks given. You should allow the committee to meet at least monthly. Meetings should be conducted according to accepted rules of order. The following is suggested as the order of business for your safety committee to follow:

- Call to order
- Roll call of members
- Introduction of visitors
- Unfinished business
- Review of accidents and injuries
- Speakers on subject of accident prevention
- Results of committee safety audits and statistics
- Promoting safety campaigns
- New business
- Adjournment

Minutes of the meetings should be taken and circulated. Committee members should be given tasks and should be present when safety awards are given to employees. You should show your appreciation to committee members for their service, especially when their ideas result in less injuries.

SUCCESSFUL SAFETY MEETING

Making a safety meeting successful takes a team of people who actively exchange ideas to accomplish goals. To help ensure success, the members should understand the purpose of the meeting. A lot can be accomplished when the members have a clear understanding of what needs to be done. Some common goals are

- Exchange information
- Solve problems
- Share concerns
- Make decisions

The members should be prepared, prior to the meeting, in other words, they should do their homework. A smart chairman will solicit questions from the members, because some members may be shy at first and that is a good way to get members to open up. Members should express their feelings. Differences of opinion expose members to other points of view. Members should keep an open mind when a member brings up an idea.

SUMMARY

Safety committees are indeed a valuable tool of management. They work well only when everyone is serious about accident prevention. I have found safety committees to be most beneficial in bringing people together to solve safety problems. Safety committees can be very effective in finding new ways to reduce accidents and injuries, and they show everyone that you are serious.

A functional safety committee can provide solutions to problems before they get out of hand and start to erode your profits.

Last, companies that invite the employee's representative to the meeting, or better, make them members, greatly benefit because no one is left in the dark when labor and management work together to reduce accidents.

17   RECORD-KEEPING AND REPORTING

"It is a good thing to learn caution by the misfortunes of others."

Syrus

LOG OF OCCUPATIONAL INJURIES AND ILLNESSES

The Occupational Safety and Health Act of 1970 requires employers to maintain records of occupational injuries and illnesses as necessary or appropriate for enforcement of the Act. The records are also required for developing information regarding the causes and prevention of accidents and illnesses, and for maintaining a program of collection, compilation, and analysis of occupational safety and health statistics.

The Code of Federal Regulations, Title 29, Part 1904.2, states that each employer shall maintain in each establishment a log and summary of all recordable occupational injuries and illnesses for that establishment. The employer shall also enter each recordable injury and illnesses in the log and summary as early as practicable, but no later than six working days after receiving information that a recordable injury or illness has occurred. For that purpose, OSHA Form Number 200 or an equivalent form shall be used.

SUPPLEMENTARY RECORDS

In addition, as an employer you must also maintain a supplementary record for each occupational injury or illness and it shall be completed in the detail prescribed in the instructions accompanying OSHA Form Number 101. Alternative records such as workers' compensation, insurance, or other reports are acceptable, providing they contain the information required by the above form.

ANNUAL SUMMARY

As an employer, you are required to post the annual summary of all your occupational injuries and illnesses. OSHA Form Number 200 shall be used in presenting the summary.

Also, you will certify that the annual summary is true and accurate by your signature. The form will be posted no later than February 1, and shall remain in place until March 1. It should be posted in a conspicuous place where it can be seen by all of your employees.

RETENTION OF RECORDS

All the records of all occupational injuries and illnesses shall be maintained for 5 years, following the end of the year to which they relate. A good point to remember is, the log and summary of all recordable occupational injuries and illnesses must, upon request, be made available to any employee, former employee, and to their representatives for examination and copying.

REPORTING OF FATALITY OR MULTIPLE HOSPITALIZATIONS

Within 48 hours after the occurrence of a fatal accident to an employee or hospitalization of five or more employees, the employer must report the accident either orally or in writing to the nearest office of the Area Director of the Occupational Safety and Health Administration (OSHA), U.S. Department of Labor.

GOOD RECORDS CAN HELP

Now at this point, you are probably thinking, "How can these records help my company or firm?" With good records, you can perform a statistical analysis of your injuries and illnesses. You can also compare your firm's safety record with the record of other firms. All you need is the number of hours your employees worked and the number of injuries and illnesses your firm experienced.

Computing incidence rates show the equivalent number of injuries and illnesses per 100 full-time employees. These rates can help you determine both problem areas, or your progress in preventing work-related injuries and illnesses.

HOW TO COMPUTE AN INCIDENCE RATE

You can compute an incidence rate of injuries and illnesses for your firm quickly and easily. The formula requires:

1. The number of injuries and illnesses in a given year. You can do this by counting the number of line entries in your OSHA log number 200.
2. The number of hours *all* employees actually worked in the same given year (do not include vacations, sick leave, holidays, etc.) An incidence rate of injuries and illnesses may now be computed from the following formula:

$$\frac{\text{Number of injuries and illnesses} \times 200,000}{\text{Employee hours worked}} = \text{Incidence rate}$$

The 200,000 hours in the formula represents the equivalent of 100 employees working 40 hours per week, 50 weeks per year. Example: Your firm experienced 11 recordable injuries and illnesses during a given year. The total hours worked by all employees during this period were 130,000 hours:

$$\frac{11 \times 200,000}{130,000} = 16.9 = \text{Incidence rate}$$

Therefore, your firm experienced a rate of 16.9 injuries and illnesses per 100 full-time employees during the given year.

COMPUTE INCIDENCE RATE FOR PREVIOUS YEARS

You may also be interested in computing incidence rates either for previous years to see how your rate has changed, or by department, or on a monthly basis in order to measure the effectiveness of a new safety campaign or other factors. The formula remains the same. Be sure, however, that the number of injuries and illnesses corresponds to the hours worked for the department or time frame you are using. Always remember that past performance is the best guide to improve your safety and health program.

Reviewing your records may help to identify characteristics common to several cases. For example, several injuries may have resulted from the use of a particular type of equipment. Once an unsafe condition or act is identified as a result of an inspection or accident investigation, you can take steps to eliminate that problem.

Yes, record-keeping is not only a requirement of the OSHA Act, but it makes good "cents" for your business.

Make certain that whoever computes your rates is not doing it to please you, or is half-hearted about doing it. I am reminded that Disraeli once said, "There are three kinds of lies: lies, damned lies, and statistics."

BUREAU OF LABOR STATISTICS (BLS)

The Bureau of Labor Statistics in the Department of Labor obtains the logs for some 275,000 workplaces each year. They then compile the information received into an annual survey report. This report provides detail information on injury and illness rates. The rates are given to OSHA for purposes of targeting certain industries and establishments for inspection.

The basic policy for many years has been to schedule an inspection of those industries and establishments that show the greatest number of injuries. When an OSHA inspector arrives at your workplace, the first request will be to examine your OSHA log 200 (daily log of injuries and illnesses). If your rate is below the Bureau of Labor's rate for your industry, the inspection will be terminated at this point.

SUMMARY

You have no choice regarding record-keeping — it is the law. However, you can use the record-keeping requirement to your advantage. Because you also need to know whether or not you are out-of-control with your injuries and illnesses, and to pinpoint the areas that are experiencing too many injuries — it will make good sense to keep good records of your accident losses.

Also, record-keeping and statistical information can be very useful to help you determine which work group is achieving good safety records. You can also compare the productivity of your work groups with their safety records. Remember that when you do this, your workers will know that you are serious about safety, and also you will be saving money because accidents and injuries are very costly.

Rewarding the worker for working safely is visible evidence that you consider your workers to be your most important asset.

Remember, OSHA requires that all employers with 11 or more employees must keep records of work-related injuries and illnesses.

18   ACCIDENT COST

"Business will be either better or worse."

Calvin Coolidge

THE MATHEMATICS OF ACCIDENTS

(+) Adds to your troubles
(−) Subtracts from your profits
(×) Multiplies your losses
(÷) Divides the worker from the manager
(%) Discounts your successes

ACCIDENTS ARE COSTLY

The prime objective of your company is to produce a product and make money. Sometimes, trying to make your company a safe place comes in conflict with your first objective. However, studies have shown that accidents decrease profits; on the other hand, spending money to make your workplace safer will decrease your losses and increase your profits.

Truly, a safe workplace is an efficient workplace and vice versa. It is an accepted fact that safety and efficiency go hand in hand. There exists a high degree of correlation between safety and productivity; the combination of low accident rates and high production rates does exist in this country's most successful businesses.

Think about this, every accident suffered by an employee can eat away at your profits. To offset even so small a loss as one caused by a $500 work accident, a business must produce at least that amount in *profits*. In other words, if your company makes a 10% profit on every $1 transaction it handles, your business must handle 5,000 transactions just to break even. Of course, most disabling work injuries cost far more than the $500 used in the above example.

No matter how you slice it, accidents do cut into your company's profits. Relating these accident losses to your company's gross receipts quickly establishes in your mind that accidents are very costly in both direct and indirect costs.

DIRECT AND INDIRECT COSTS

While it is recognized that some companies only look at direct costs, their greater losses are the indirect costs. Direct costs vs. indirect costs can be compared to an iceberg. One third of the iceberg is above water (direct), while two thirds is below the top of the water where it is not seen (indirect).

Like the part of the iceberg you cannot see, the indirect cost of an accident is not readily visible unless you consciously think about the hidden cost of each accident.

I have listed some direct costs and some indirect costs. I am sure that your accounting department can come up with other costs to add to the two lists. Every accident that you investigate will provide you with cost estimates that will be different from accident to accident. You may want to average out many accidents to arrive at an average cost for each direct and indirect charge.

EXAMPLES OF DIRECT AND INDIRECT COSTS

Direct Costs

- First aid and emergency transportation
- Medical services
- Continuation of wages to the injured worker
- Compensation payments

Indirect Costs

- Time spent in accident investigation
- Preparation of accident/injury forms
- Interruption or stoppage of work
- Idleness of onlookers
- Low morale and decline in job performance
- Repair or replacement of material/equipment
- Loss of productive effort
- Extra cost for overtime necessitated by the accident
- Cost of hiring new employees
- Cost of learning time for new workers
- Rework time on spoiled products
- Increased insurance premiums
- Reduced competitive position from late delivery
- Civil penalties, fines, legal fees, and lawsuits

OTHER LOSSES

Many other losses will be incurred when an injured worker with a bandaged hand will not be able to handle machinery or material in the same way as before the accident. Realizing that out of sympathy, you might give the worker a light duty job or even create a temporary job, this will further erode your profits.

At this point, you might ask, "What is the cost of an accident in terms of direct and indirect losses?" My answer is, "There is no set cost because industries and projects are not structured the same way." The losses incurred by one company would be different in another company.

You should consider conducting a pilot study where cost figures are assigned to each direct and indirect losses. This will allow you to determine average uninsured costs per accident case. Many successful companies who have done this study have greatly reduced their losses by learning where to fix the problems and have thus remained competitive or ahead of other industries.

COST YOUR ACCIDENTS

Below is a list of questions regarding an accident that may have happened in your plant or project at some time or other. Answer the questions by putting an estimated cost for each question, then total all costs, and that will be about what an accident will cost you. Do not toil over the accuracy of each cost because you will most likely arrive at low estimates. The estimate, even with a low figure, will change your mind about accidents and injuries and how they take away from your profits.

THE COST OF AN ACCIDENT

- When the accident happened, the injured employee stopped
  working. How long was the employee away from work? .... $_____
- Your first aid or medical group treated the injured worker
  and prepared an accident report. How much time did it take? ...... $_____
- The supervisor and others conducted an investigation of the
  accident and prepared the forms as required by OSHA.
  How much time was spent in investigating the accident
  and typing reports? .... $_____
- Other workers stopped their work or machines. How long
  were they stopped? .... $_____
- Witnesses were questioned by the supervisor. How much
  time did that take? .... $_____
- Were there any tools or equipment damage as a result
  of the accident? What was the cost of replacement
  or repairs? .... $_____

- Were the injured person's duties taken over temporarily by another nearby employee? What was the cost to keep production going? .... $_____
- If the accident required you to hire a new employee, what were the hiring costs? .... $_____
- What was the cost to train the new employee to do the same job? .... $_____
- If the product was spoiled, what was the cost of rework? .... $_____
- If the accident slowed production, what was the cost for overtime necessitated by the accident? .... $_____
- If the worker was sent home to recuperate, how much did you pay in continuation of wages? .... $_____
- If OSHA found you negligent, how much were you fined in penalties? .... $_____
- If your insurance payments are increased because of the accident, how much? .... $_____
- If legal proceedings are brought against you, how much will this cost? .... $_____
- If the employee returns to work, but under medical advisement is unable to work at his/her own job, and you assign that employee to a temporary job, what does this cost you? .... $_____
- Add any other expenses not listed here and total the lot. .... $_____

Do you now believe that accidents are expensive? What is the permanent loss of a skilled employee worth to your organization? Or worse, what is a worker's life worth? When you have answered to yourself, too much, you are on your way to improving your safety program.

SUMMARY

You must believe that accidents, injuries, fires and other disasters are costly to you, then you will be on your way to increasing your profits.

No company or business can exist for long unless immediate action is taken to reverse an unacceptable accident trend. To do nothing is to invite trouble.

Take a good look at your accident costs. They are losses that slip out the back door of your company while you are watching your front door. Look at your accident/injury statistics, then go down to the work areas and take a good look — after all, it is your money! Share the problem with your managers and supervisors. You can also put your safety committee to work on a solution. Whatever you do, do not ignore the high cost of accidents — it is an indicator that things will get worse if left alone. And, never forget that OSHA is always ready to do a wall-to-wall inspection if your injury incidence rate is above the national average.

19 INCENTIVE AWARDS

"Lack of something to feel important about is almost the greatest tragedy a man may have."

Dr. Arthur E. Morgan

REWARDING EMPLOYEES

People work because they expect something from it. The same applies when people are asked to perform their work in a safe manner — they want recognition! They also feel one of two ways about safety, some people want to be safe, while others are chance takers who deviate from your safe work practices.

The workers who want to be safe are self-motivated to safety and will comply with your safety policy, while the others will take shortcuts in order to get the job done faster, but sooner or later they will suffer an accident that will be very costly. Why? Because they wanted to be recognized as fast workers at the expense of an injury. Ignoring safety rules for the sake of more production is not only foolish, but it is economically unsound.

To begin with, people expect a reward when they perform their work to your highest level of expectation, so, why not combine productivity and safety together? In other words, the reward should be given to those high achievers who performed their work without a lost-time injury.

PAYOFF OR AWARD?

There are some managers who believe that incentive awards are a payoff in order to get the worker to comply with the safety rules of a company; but whatever it is called, the payoff or incentive award can be very beneficial to the success of a company and its mission.

Some organizations provide incentives in the form of bonuses to employees who produced quality work in a timely manner, but they did not consider safety as an attribute for awarding the bonus. Including safety in the process becomes an important way of saying to the employee that the company cares about their safety. I have seen this caring for the employees "domino" throughout a business with excellent results.

RESISTANCE OF SUPERVISORS AND WORKERS

Some will argue that doing the job without an accident or injury is a condition of employment. True! However, those supervisors will never reduce their injuries by holding that attitude over the heads of their workers. Think about this, it is far better to motivate employees to working safely with incentives than to make them tow the mark as a condition of employment.

I am not saying that you should not have a company policy regarding safe work, but I am saying that you must motivate the worker to *want* to work safely; and remember that sometimes an injury to an employee is caused by another employee's unsafe act or by an unsafe condition created by the employer.

While I admit that some employees will not comply with your safety rules, nor will they support your incentive program, that of course is a "horse of a different color". However, Chapter 20, Enforcing Safety, should help you get around the workers who refuse to comply with your safety rules.

PRODUCTIVITY AND SAFETY

Traditionally, supervisors and employees are rewarded for outstanding job performance, but for the most part, not for their safety efforts. Why not combine both attributes for the basis of the award, as they both make or save you money?

You should also consider injury cost reduction sharing within each department. In other words, when your departments achieve a reduction in accident costs, you may want to share the savings with that unit. It is not a substitute for the normal and competitive wage structure, but it is built on top of it. It is also a way of making your workers feel that they are important.

Some companies invite the group with the best safety record to a nice meal and their spouses are invited to participate as well. Other companies give savings bonds to the best group, but regardless of what you do, your employees will know that you are sincere about their safety and they will come to believe that safety is part of productivity.

FINANCIAL AND NONFINANCIAL REWARDS

Many incentives have both a financial and a nonfinancial aspect. In other words, recognizing an employee that has worked a year (2,000 hours) without a lost-time injury by giving that person a certificate, belt buckle, lapel pin, etc. is considered a nonfinancial award. On the same token, giving that person a bonus or extra day vacation is considered to be a financial award.

Some companies also recognize the injury-free employee by honoring that person via the company's newsletter, bulletin board, or local newspaper. Most successful managers believe that employees will respond to incentives. As a matter of fact, the employees will brag to others about the awards and this motivates young employees to role model after outstanding workers.

Now there are some who say that safety incentive awards are expensive. I would say to them, "read Chapter 18, Accident Cost", and that should change their minds. When you begin to compare the cost of injuries, disruptions, lower morale, insurance costs, retraining, and the many, many other costs, you will agree that incentive awards are a good bargain.

Safety Award Pin

Giving a safe worker pin to your safe workers is tangible evidence that management recognizes and appreciates the individual's effort toward safety, saving lives, and equipment. This recognition, in the form of safety pins, help build morale and makes the recipient and all personnel more aware of safety factors and the importance of safety cooperation.

Safety Award Plaque

Plaques, medals, banners, etc., are usually awarded to plants, departments, sections, or groups for excellent safety and accident prevention records. Such unit awards to groups inspires the individual with a greater sense of team spirit and cooperation in safety matters and with an awareness of responsibility for the general welfare of the company.

Individual Gifts

Various types of useful gifts for deserving employees are incentives to work safely. Generally they are available with an imprinted safety motto or message. Typical items include tie clips, belt buckles, and pen/pencil sets.

Safety Award Trophy

Trophies and similar awards stimulate employee interest in company safety programs. They are given to individuals for unusual service or, more often, to groups for superior achievements in safety. A trophy is tangible evidence of management's recognition and appreciation. It helps overcome indifference and arouse the cooperative spirit.

Custom-Imprinted Apparel

Custom-imprinted clothing such as jackets, caps, T-shirts, and other wearables are available to reward safety conscious workers or to reinforce company safety programs. They emphasize the cooperation, motivation and support necessary from each employee to make safety programs succeed.

Pocket Savers

These safety pocket savers are plastic pouches or linings that fit into the pockets of shirts or jackets to protect clothing from abrasion or discoloration from pens, pencils, rulers and other pocket-clip devices. The outer flap of the pocket saver carries a safety slogan and/or safety cartoon. There is also space for imprinting the name of your company.

Plant Safety Scoreboard

An incentive aid to help promote plant safety, the safety scoreboard, is a constant reminder to personnel of the number of days worked without a lost-time injury. Provision for the use of interchangeable numbers allow daily updating of statistics. Scoreboards for outdoor use are also available.

OTHER SAFETY AIDS

There are many other ways to let your employees know that you care. One effective method that I found was the use of posters, stickers, and banners. These are available for the topics of general appeal, chemicals, personal protective equipment, electricity, falls, fire safety, housekeeping, motor transportation, tools, etc. For a complete listing of a buyer's guide, consult "Best's Safety Directory", which I listed as a useful reference after the last chapter of this book.

SUMMARY

Incentives and safety awards are strong motivators to excite people to work safely. We are all taught at a very young age to expect rewards for jobs well done. This is further reinforced in school sport programs with trophies, letters, and recognition. The military services award sharpshooter medals and other awards for outstanding achievements. Most everything we do is advertised by patches, jackets, and pins. An incentive program will inspire your employees to work safely and ultimately save you money. When you include the spouses of your employees at the awards presentation, and when you involve him or her in the benefits, you will have a person who will enforce your safety program because there is something in it for him or her.

20

ENFORCING SAFETY

"It is far better to make a worker a disciple of safety than to discipline the worker."

Author

ENFORCING SAFETY RULES

It is recognized by many managers that persuasion is preferable to compulsion. Yet, there can be no question that there must be teeth behind what is required by management. Otherwise, you can expect repeated violations. If an employee refuses to wear required protective equipment, or to follow safe work practices, then the supervisor should bring the matter to your personnel department as a way of showing to the employee that refusing to obey will not be tolerated.

POOR ATTITUDES

Your supervisors are normally in the best position to observe their workers. Indicators of attitudinal problems or problems emerging are

- *HABIT* — Absenteeism, tardiness, drug and alcohol intake
- *ATTITUDE* — Dissatisfaction, loss of interest and sensitiveness
- *EMOTION* — Edgy, excitable, strange, and depressed
- *BEHAVIOR* — Lack of confidence, poor conduct, and frequent errors

These, and many other indicators, often lead to inattention which can result in accidents, or worse, a serious injury or fatality.

Excessive or chronic absence from work and/or a series of minor injuries bear looking into. Why? Because in all probability, they are predictive that corrective action to change the employee's behavior should be taken right away. To ignore all the above is an accident waiting to happen.

COMMUNICATE SAFETY

It is very important that your supervisors communicate about safety to their subordinates in order to prevent costly accidents and injuries. And how is this done? To begin, no one is born knowing safety skills. We all have to be taught. It is a known fact that workers will follow the lead of their supervisors. When a supervisor believes in safety, the workers will follow that supervisor's lead by observing all safety rules.

Remember this, some supervisors create problems. I do not like picking on supervisors, but when supervisors do not show the right example and downplay safety, they should be made to change their attitudes toward safety, or give up the job of supervising employees. No matter how hard a supervisor tries to change a worker's attitude toward safety, it will never happen as long as the worker thinks that the supervisor is just as uncaring about observing safety rules.

SUPERVISORY CONTROLS

Your supervisors need to control the behavior of their employees. We all need controls — we get up in the morning with a control (clock). Controls are good for us and they give us purpose. When I go through a green light, I expect the other driver to stop at a red light. I also expect the other person to stop at a stop sign. I cannot imagine a society without controls — what a mess that would be.

It will not help your safety program if workers see, that in practice, management often tolerates safety infractions. When your supervisors see their workers breaking safety rules, they should nip these right in the bud. To overlook those violations can cause your safety program to go downhill in a hurry, and the impact of this "overlook" can be very costly.

One way I know of that will defeat your supervisors ability to communicate safety to employees is when they discriminate between workers. In other words, they initiate corrective action for unsafe behavior of some workers, while they overlook the same poor behavior in others. Your supervisors should treat each infraction and they should let their workers know in advance that action will be taken as soon as possible when employees are found breaking safety rules and safe work practices.

Supervisors should take every opportunity to instruct their workers in job safety. They should never put off correcting a bad habit when it is observed. No worker wants to hear that they were working in an unsafe manner several days ago. Do not wait, take action right away to correct unsafe behavior.

WARNING THE EMPLOYEE

If the supervisor has exposed the unresponsive worker to safety training, stand-up safety talks, and has shown a good example, then the only course of action left is to invoke the disciplinary process. The supervisor should make sure that the worker knows, in advance, the consequences of breaking safety rules. Along with this, supervisors should let their workers know every day that they care and are concerned about their safety.

Supervisors can jump on the workers for little things, but when they do this, the workers will duck the supervisors and will never tell them of unsafe conditions. When this happens, the supervisor will never know what is really happening until it is too late. The key is to be reasonable and consistent. Some supervisors I have known have tried manipulative reciprocity with their workers, "you do something for me, and I'll do something for you." That can backfire and give anyone who tries this a really bad problem. You can't promise everything to everyone!

DISCIPLINARY ACTION

The real key to changing behavior, or in invoking disciplinary action is to be reasonable, up front, do not jump to conclusions, and be fair. Your supervisors have three things going for them:

- Power — supervisors have the backing of the company.
- Authority — supervisors are in charge of employees.
- Right — supervisors have the right to take corrective action.

Take away any one of the above, and your supervisors are rendered ineffective. This could cause you problems and loss of productive effort.

ANOTHER WAY

Instead of involving disciplinary action, let your supervisors know that changing the workers' behavior is the better way to effect a change. You do not want to terminate an employee, unless they absolutely refuse to comply with safety rules. Remember that the first action taken by your supervisor in a timely manner might very well correct a worker's behavior. On the other hand, the lack of proper and timely action to correct poor behavior, could cost you much money; especially when it can be proven at a later date that the supervisor was lax in enforcing safety rules. Behavior modification can create a more productive climate, build better morale, and help establish trust between workers and supervisors.

PERFORMING AN AUDIT

I have seen many companies with very low accident rates, and that did not happen by chance. Instead, what I saw was a climate in which supervisors and workers knew in advance what to expect from each other. It was also an understanding that the worker knew what to expect from the supervisor in terms of disciplinary action. When you want to know how effective your supervisors are in communicating safety to your employees, take a hard and long look at your log of accident and injury statistics within each supervisory group. Then go down to the work area and take another good look — you will be surprised. The supervisor who is lax towards safety will also be ineffective regarding housekeeping. Next, you will see that the employees of that same supervisor will be found ignoring your safety rules.

BEHAVIOR MODIFICATION

Disciplinary action should be the last resort. This is not to say that a worker should not be disciplined when found breaking your safety rules. Instead, there are a number of actions your supervisors can take with a worker — actions that will instill positive behavior regarding safety rules. Probably the best action is motivation. Motivation is a widely used method in directing and controlling the activities and behavior of people. Motivation is particularly important in accident prevention because of the success achieved by a number of companies in reducing their accidents and injuries to an acceptable level. Most companies are sincerely interested in the safety of their employees. They do not want people injured, nor do they want to see the worker's family suffer because of an injury.

SUMMARY

Bear in mind, it is very expensive to replace a worker because no one took time to change the worker's poor behavior. It is doubly expensive when the employee was discharged unfairly. Especially when the terminated worker appeals the action taken by you, and wins the appeal on grounds that the supervisor did not care. It is always better to change the worker's behavior before taking the last resort. The courts will be more sympathetic to you, when they know you have tried to correct unsatisfactory behavior. Most of the time, the real problem can be assigned to lack of communication. As Maribeau said, "It is always a great mistake to command when you are not sure you will be obeyed."

21 FIRST AID AND MEDICAL PERSONNEL

"All is to be feared when all is to be lost."

Byron

First aid means just that — it is the first treatment you give to an injured person. The Code of Federal Regulations, OSHA, Title 29, Part 1910.151(a), requires that each employer shall ensure the ready availability of medical personnel to advise and consult on matters relating to workplace health.

OSHA has defined "medical personnel" to mean a physician with a valid license to practice medicine. In order to advise and consult, the physician needs to know how your company operates insofar as your work processes are concerned and the hazards associated with what you do. In other words, how you do your work, where you do your work, the work environment, and the material or equipment required to accomplish each task. If you are using hazardous substances, the physician will need to know that. (See Chapter 11, Hazard Communication).

MEDICAL PERSONNEL

OSHA, Title 29, Part 1910.151(b), deals with the absence of a clinic, hospital, or infirmary. In some cases, the workplace will not be in close proximity to a clinic, infirmary, or hospital, and the workplace will not have available a full-time nurse or doctor on duty. When this is the case, at least one, but preferably two or more persons per shift should be trained to give first aid treatment. Also they should have a current certificate from an approved first aid training program, such as:

- American Red Cross
- American Petroleum Institute First Aid Training Program
- Mine Enforcement and Safety Administration First Aid Training Program
- Other courses approved by the OSHA Regional Administrator

Federal courts define "close proximity" to mean 3 to 4 minutes from a medical facility. Accidents resulting in severe bleeding or respiratory arrest can be fatal beyond that time. First aid supplies must be available and your consulting medical people must approve the first aid supplies. The supplies should be appropriate in type and quantity for the type of operation performed at your workplace.

EYEWASH AND DRENCHING SHOWER

OSHA Title 29, Part 1910.151(c), specifies that facilities for quick drenching and flushing of eyes and body must be provided in the work area if injurious corrosive materials, for instance, strong acids, are present. Drenching or flushing means to let water run over a person. Drenching and flushing facilities include showers and eye wash fountains, and in certain applications, neutralizing solutions in bottles.

Whenever flushing is used, it should be for at least 15 minutes, and then the employee should be taken to a hospital by ambulance. The ambulance crew will continue flushing the patient until they arrive at the hospital where the transfer for care is accomplished.

In permanent workplaces when the extent of the hazards so dictates, the drenching (shower) and the flushing (eyewash) facilities should be connected to a permanent water system. However, in temporary or remote locations, portable eye wash fountains or portable showers are acceptable, depending upon the extent or magnitude of the hazard.

All drenching and flushing units should be located in close proximity to the hazard. That is, the distance from any given potential hazard to the shower and eye wash station should not exceed 25 feet in walking distance. The first 15 seconds after injury is critical, so none of this time should be spent looking for an eye wash bath or safety shower, even in the immediate area.

It is important to note that hand-held drench hoses support shower and eye wash units, but they do not replace them. For more information concerning the above requirements, consult your local safety equipment center.

CARDIOPULMONARY RESUSCITATION

Cardiopulmonary resuscitation (CPR), is the combination of artificial respiration and manual artificial circulation that is recommended for use in cases of cardiac arrest. CPR provides proper care at the earliest possible time, which is vital. Cardiopulmonary resuscitation involves the following "ABCD" steps:

- Airway opening
- Breathing restored
- Circulation restored
- Definitive therapy

Your workers should know how to perform CPR because this could help to save a life, especially when it could involve a worker receiving an electric shock or a worker that might be involved in a process that could suffocate him or her. CPR is also a life saver in cases of heart attack. Your local American Red Cross can assist you in providing training for your workers. Cardiopulmonary resuscitation can make the difference between death and many years of useful service.

You can start by training your employees who are involved in life-threatening job situations. For example:

- Working with electricity
- Working in confined spaces
- Working with chemicals
- Working aloft

SETTING UP A FIRST AID STATION

The first step in setting up your plant's first aid program is the appointment of an individual who will be responsible for the entire program. This person could be your safety manager, or the plant's nurse, who could also maintain your injury log, record-keeping, and statistics. You should set aside a first aid room for treating ill or injured workers.

If your business is small and you do not have a full-time nurse, first aid kits should be located throughout the plant so that no employee is more than a few minutes away from first aid. If you elect to do the latter, each kit must be under the supervision of an assigned individual, usually the qualified first aid person.

To run a successful first aid program, your designated first aid person must keep good records of each treatment (see Chapter 17, Record-Keeping and Reporting).

OSHA 29 CFR, Part 1910.410 requires that all dive team members shall be trained in CPR and first aid, as approved by Red Cross standards or an equivalent course of study. The emergency first aid and first aid supplies for the above are specified in OSHA 29 CFR, Part 1910.421.

REQUIRED PHYSICAL EXAMINATION

Workers who are required to wear respirators must be evaluated by a physician prior to wearing a respirator. Those workers should be re-evaluated annually (OSHA 29 CFR, Part 1910.134).

Medical examination and surveillance is required for the hazards listed in OSHA 29 CFR, Part 1910.1001 through 1910.1047. The medical evaluation will determine if the workers are able to function while wearing a respirator.

CONSTRUCTION WORK

OSHA 29 CFR, Part 1926.803, requires physical examinations for construction employees who must work in compressed air. The regulation also specifies the facilities required to treat the decompression illness of divers. Medical service and first aid for construction employees is specified in OSHA 29 CFR, Part 1926.50.

Drenching or flushing units to flush battery acid from the eyes and body is a requirement of OSHA 29 CFR, Part 1926.403.

There are other requirements of OSHA in 29 CFR, Part 1910 through Part 1926 that will specify when physical examinations are required. If you are unsure of what is required, obtain a copy of the regulation that applies to your business.

SUMMARY

The foremost objective of your company's first aid program must be to protect the life of an employee in the event of a severe injury. Your company should make an effort to reduce minor injuries to a level as low as possible. Often minor injuries such as cuts, bruises, burns, etc., which, when not treated immediately, can result in more serious injuries.

Many types of first aid kits/cabinets are available from your local first aid equipment supplier. Remember that even the finest first aid products are only as effective as the individual using them. Your company should have employees trained in emergency first aid techniques.

If you look around, you may find that some of your employees may have received this training and are presently serving their communities as first aid volunteers with local emergency units.

Last, always treat your first aid program as an investment that is designed to save your company's assets, and more importantly, your people.

22 OSHA

"Ignorance never settles a question."

Disraeli

The William-Steiger Occupational Safety and Health Act of 1970 requires that every employer under the Act, shall provide to his employees, a place of employment which is free from recognized hazards, that are causing or are likely to cause death or serious physical harm to employees. The Act also requires that employers comply with the Occupational Safety and Health standards of the Act, and that employees shall comply with standards, rules, regulations, and orders issued under the Act.

OSHA STANDARDS

OSHA standards fall into four major categories:

- General industry
- Maritime
- Construction
- Agriculture

The above standards, or the Federal Register, which is one of the best sources of information on the above standards, is available in many public libraries. Annual subscriptions are available from the Superintendent of Documents, U.S. Government Printing Office, Washington, DC 20402. For more information, contact your nearest OSHA office.

INFORMING YOUR EMPLOYEES

A job safety and health protection workplace poster (OSHA 2203), informing employees of their rights and responsibilities under the Act is also available from your nearest OSHA office. The poster must be displayed in a conspicuous place where it can be seen and read by all your employees. Beside displaying the workplace poster, the employer must make copies of the Act and copies of relevant OSHA rules and regulations available to employees upon request.

Also, you must post a copy of your petition to OSHA for variances from standards or record-keeping procedures, copies of all OSHA citations for violation of standards, and the annual summary log of your occupational injuries and illnesses (OSHA 200), which must be posted before February 1, and must remain posted until March 1, each year.

All employees have the right to examine any records kept by their employers regarding exposure to hazardous materials, or the results of medical surveillance and special testing and/or workplace monitoring results.

OSHA INSPECTIONS

The OSHA Act is enforced by compliance officers who are authorized by law to enter and inspect your plant or project. Based on the compliance officer's findings, the OSHA area director may issue citations for alleged standards violations and designate a period of time by which the alleged violations must be abated (fixed).

When the OSHA compliance officer arrives at your workplace, be sure to carefully examine his credentials. There have been instances where industrial spies posing as OSHA inspectors have entered a competitor's plant. Other imposters have attempted to collect money for penalties or sell "OSHA approved" equipment to plant management. When in doubt, call the nearest OSHA office to verify the inspector's credentials.

If you are satisfied that the inspector's credentials are legitimate, you should allow him or her to begin the inspection. You can refuse to let the OSHA inspector enter, but you either "pay now or pay later". The OSHA inspector will explain the scope of the inspection to you. It could be the result of too many accidents, an employee complaint, or worse, a fatality.

Once you consent to admit an OSHA compliance office into your workplace, you cannot later claim that the inspection constituted a warrantless search that violated your Constitutional rights under the Fourth Amendment. Personally, I think it is foolish to refuse OSHA entry into your workplace. They will not get mad, but they will get even.

With very few exceptions, OSHA inspections are conducted without advance notice. In fact, alerting an employer in advance of an OSHA inspection can bring a fine of up to $1,000 and/or a 6-month jail term. There are, however, special circumstances under which OSHA may indeed give notice to the employer, but even then, such a notice will be less than 24 hours. These special circumstances include:

- Imminent danger situations which require correction as soon as possible
- Inspections which must take place after regular business hours, or which require special preparation
- Cases where notice is required to assure that the employer and employee representative or other personnel will be present
- Situations in which the OSHA area director determines that advance notice would produce a more thorough or effective inspection

Employers receiving advance notice of any inspection must inform their employees' representative (union), or arrange for OSHA to do so. If an employer refuses to admit an OSHA compliance officer, or if an employer attempts to interfere with the inspection, the act permits appropriate legal action.

INSPECTION PRIORITIES

Obviously, not all 5 million workplaces covered by the Act can be inspected immediately. The worst situations need attention first. Therefore, OSHA has established a system of inspection priorities:

- Imminent danger situation — where there is reasonable certainty that a danger exists that can be expected to cause death or serious physical harm immediately. Health hazards are also considered as imminent danger. The compliance officer will ask the employer to voluntarily abate the hazard, and to remove employees from exposure. Walking off the job because of potentially unsafe workplace conditions is not ordinarily an employee right. To do so, may result in disciplinary action by the employer. However, an employee does have the right to refuse (in good faith) to be exposed to imminent danger.
- Catastrophes and fatal accidents are given second priority, to investigate fatalities and catastrophies resulting in hospitalization of five or more employees. Such situations must be reported to OSHA by the employer within 48 hours. Investigations are made to determine if OSHA standards were violated and to avoid recurrence of similar accidents.
- Employee complaints of alleged violations of standards or of unsafe or unhealthful working conditions are given third priority. The Act gives each employee the right to request an OSHA inspection when they feel that imminent danger exists, or that a violation of an OSHA standard threatens physical harm. OSHA will maintain confidentiality if requested by the employee.
- Next in priority are programs of inspection aimed at specific high-hazard industries, occupations or health substances. Industries are selected for inspections on the basis of such factors as the death, injury, and illness incidence rates and employee exposure to toxic substances. Comprehensive safety inspections in manufacturing will be conducted only in those establishments with lost work day injury rates at or above the most recently published national rate for manufacturing.

OSHA will also conduct follow-up inspections to determine if the previously cited violations have been corrected.

Upon arrival of the OSHA compliance officer to your workplace, the OSHA inspector will request an opening conference with you and the employee representative. The employee representative will also accompany the compliance officer and yourself during the inspection of your workplace. The compliance officer, during his inspection tour, will consult with employees, may take photos, may take instrument readings, and examine records. Trade secrets observed by the compliance officer must and will be kept confidential.

Some apparent violations detected by the compliance officer can be corrected immediately. When they are corrected on the spot, the compliance officer will note the employer's good faith in complying with the act.

CLOSING CONFERENCE

After the inspection tour, a closing conference is held between the compliance officer and the employer. It is a time for free discussion of problems and needs; a time for frank questions and answers. The compliance officer will discuss all unsafe or unhealthful conditions observed during the inspection and all apparent violations for which a citation may be issued or recommended.

The employer is told of appeal rights. After having received and reviewed the report, the OSHA area director will indicate proposed penalties. Bear in mind that the way you conduct yourself with the compliance officer is, at times, directly proportional to the fine or penalty imposed. A state of cooperativeness is strongly encouraged by this writer.

OSHA CITATIONS AND PENALTIES*

Other Than Serious Violation — A violation that has a direct relationship to job safety and health, but probably would not cause death or serious physical harm. A proposed penalty of up to $1,000 for each violation is discretionary. A penalty for any other serious violation may be adjusted downward by as much as 80%, depending on the employer's good faith (demonstrated effort to comply with the Act), history of previous violations, and size of business. When the adjusted penalty amounts to less than 60%, no penalty is proposed.

Serious Violation — A violation where there is a substantial probability that death or serious physical harm could result, and that the employer knew, or should have known, of the hazard. A mandatory penalty of up to $7,000 for each violation is proposed. A penalty for a serious violation may be adjusted downward, based on the employer's good faith, history of previous violations, the gravity of the alleged violation and size of business.

---

* All monetary penalties cited are subject to change, therefore it is wise to get this information from OSHA on an annual basis.

Willful Violation — A violation that the employer intentionally and knowingly commits. The employer either knows that what he or she is doing constitutes a violation, or is aware that a hazardous condition existed and made no reasonable effort to eliminate it. Penalties of up to $70,000 may be proposed for each willful violation. A proposed penalty for a willful violation may be adjusted downward, depending on the size of the business and its history of previous violations. Usually, no credit is given for good faith.

If an employer is convicted of a willful violation of a standard that has resulted in the death of an employee, the offense is punishable by a court imposed fine of $250,000, or by imprisonment for up to 6 months, or both. A second conviction doubles these maximum penalties.

Repeated Violation — A violation of any standard, regulation, rule or order where, upon reinspection, a substantially similar violation is found. Repeated violations can bring a fine of up to $70,000 for each such violation. To be the basis of a repeat citation, the original citation must be final; a citation under contest may not serve as the basis for a subsequent repeat citation.

Failure To Correct Prior Violation — Failure to correct a prior violation may bring a civil penalty of up to $1,000 for each day the violation continues beyond the prescribed abatement time.

Falsifying Records — Falsifying records, reports, or applications upon conviction can bring a fine of $10,000 or up to 6 months in jail, or both.

Posting Violation — Violations of posting requirements can bring a civil penalty of up to $3,000.

Assault — Assaulting a compliance officer, or otherwise resisting, opposing, intimidating or interfering with a compliance officer in the performance of his or her duties is a criminal offense, subject to a fine of not more than $5,000 and imprisonment for not more than 3 years.

Citations and penalty procedures may differ somewhat in states with their own Occupational Safety and Health programs.

LEGAL TERMS

- **CARE** — That degree of care exercised by a prudent person in observance of legal duties toward others.
- **NEGLIGENCE** — Failure to exercise a reasonable amount of care toward others.
- **TORT** — A wrongful act or failure to exercise care.
- **LIABILITY** — an obligation to rectify or recompense any injury or damage.

SUMMARY

The Occupational Safety and Health Act of 1970 was primarily designed to protect the employee. However, as time has passed, the Act has also protected the employer — at a time when science and industry was seen to be moving ahead by leaps and bounds.

I have seen employers within the last few years calling OSHA for assistance. I have also observed that the typical present OSHA compliance officer has matured greatly compared with the compliance officer of the early 1970s. The officers are courteous, tactful, and willing to help the employer to comply with the regulations.

The companies that refused to change are now out of business because of their high injury rates and subsequent OSHA inspections. OSHA is funded by the taxpayer, therefore, as a taxpayer, you should get your money's worth. There are many ways that OSHA can be very helpful in reducing your accident losses. Whatever you do, do not fight them or ignore the OSHA regulations — it can be very costly.

23   DRUGS AND ALCOHOL

"Vice repeated is like the wandering wind; blows dust in others eyes, to spread itself."

Shakespeare

DRUG AND ALCOHOL INTAKE

The United States is a drug-oriented society. We gulp pills to sleep, to take away pain, to gain or lose weight, and to get high. Add all this to the many drugs prescribed by physicians and we have a problem. Also, there are 95 million alcohol consumers in the United States who drink with meals, sports, friends, and at parties.

It is a fact that drinking alcoholic beverages is generally acceptable in our country today as part of our way of life. The combination of drugs and alcohol can produce a variety of effects that may severely impair a worker. Concentrations of drugs and alcohol remain in the bloodstream much longer than most users realize, and the effects of this combination may produce some undesired results. There is sufficient evidence to conclude that such a combination can lead to impairment of judgment and skill.

THE USER OR ABUSER

Drugs have a tendency to distort or block a worker's decision-making capability. Managers and supervisors should be alert to employees whose work performances are deteriorating because of intake. These employees should be referred to medical care.

The following list of drugs will help you understand the types:

- Alcohol (alcoholic beverages)
- Cannabis (marijuana)
- Depressants (sedatives-hypnotics)
- Hallucinogens

- Inhalants
- Narcotics
- Stimulants
- Tranquilizers

Alcohol

Alcohol is a depressant affecting first the higher reasoning areas of the brain, with perhaps a feeling of relaxation or, in the company of others, a sense of exhilaration and conviviality due to the release of inhibitions. Later, motor activity, motor skills, and coordination are disrupted, and with deepening intoxication, other bodily processes are disturbed. The drinker is often unaware of detriment to his or her normal skills and should be restrained from activity requiring such skills, particularly driving and operating machines or equipment.

Cannabis (Marijuana)

Cannabis sativa is an herbaceous annual plant that grows wild in temperate climates in many parts of the world. Marijuana usually consists of crushed cannabis leaves and flowers, and often twigs. Hashish is a preparation of cannabis resin and is generally five or more times as potent as marijuana. Marijuana is smoked; hashish may be smoked, but is also commonly made into a confection or beverage.

The immediate physical effects of smoking one or more marijuana cigarettes include:

- Throat irritation
- Increased heart rate
- Reddening of the eyes
- Occasional dizziness, lack of coordination, or sleepiness
- Increased appetite

Among the effects are feelings of exhilaration, hilarity, and conviviality; but there is also distortion of time and space perception, and there may be disturbance of psychomotor activity, which could impair driving and other skills.

Depressants (Sedatives-Hypnotics)

Depressants (downers) are drugs that act on the nervous system, promoting relaxation and sleep. Chief among these drugs are the barbiturates, the more important of which are

- Phenobarbital (goofballs)
- Pentobarbital (yellow jackets)

- Amobarbital (blue devils)
- Secobarbital (red devils)
- Methaqualone (quaaludes)

Others closely related are

- Glutethimide (doriden)
- Chloralhydrate (knockout drops)
- Paraldehyde
- Other depressants (Ativan, Dalmane, Placidyl, Restoril, and Valmid)

A usual therapeutic dose of a barbiturate does not relieve pain, but has a calming relaxing effect that promotes sleep. Reactions include:

- Relief of anxiety and excitement
- Tendency to reduce mental and physical activity
- Slight decrease in breathing

Barbiturates are used to control convulsions. An overdose can produce unconsciousness, deepening to a coma.

Hallucinogens

Hallucinogens are drugs that are capable of producing mood changes, frequently of a bizarre character; disturbance of sensation, thought, emotion, and self-awareness; alteration of time and space perception; and both illusions and delusions. The most important hallucinogen is Lysergic Acid Diethylamide (LSD). Some others are

- Mescaline
- Psilocybin — PSILOCYN
- Semylan — PCP
- Morning Glory Seeds
- A number of synthetic substances

Abuse of hallucinogens is of the spree type. The drug is taken intermittently, although perhaps as often as several times a week. LSD, for example, is likely to produce these physical effects:

- Increased activity on the central nervous system
- Increased heart rate
- Increased blood pressure
- Increased body temperature
- Dilated (enlarged) pupils
- Flushed face

The effects are highly variable and unpredictable. Ordinary things appear beautiful, colors seem to be heard, and a mood of joy and peace may also mark the use of the hallucinogens. Some undesirable effects are: a complete loss of emotional control, profound depression, tension, and anxiety.

Inhalants

Occasional self-administration of volatile substances such as ether or chloroform in order to experience intoxication is a very old practice. In recent years, inhalation of a wide variety of substances, a practice commonly referred to as glue-sniffing, has become widespread among young people in their early teens. The substances inhaled include:

- Fast-drying glue or cement
- Many paints and lacquers and their thinners and removers
- Gasoline
- Kerosene
- Lighter fluid and dry cleaning fluid
- Nail polish and remover

The usual method of inhaling is to hold a cloth over the nose and mouth with some of the substance on it or to cover the head with a paper or plastic bag containing a quantity of the substance. The effects resulting from the use of inhalants are those experienced in abuse. Reactions are

- Initial excitement resulting from release of inhibitions
- Irritation of the respiratory passages
- Unsteadiness
- Drunkenness, with growing depression that deepens even to unconsciousness

The person may suffocate before removing the bag from the head, and some of the propellants in the aerosols that are inhaled are toxic to the heart and can cause death by changing the rhythm of the heartbeat.

Narcotics

The term narcotics refers, in general, to opium and specifically to:

- Preparations of opium, such as paregoric
- Substances found in opium (morphine and codeine)
- Substances derived from morphine (heroin, dilaudid, etc.)
- Synthetic substances like merperdine, demerol, methadone, methadose, or dolopphine.

A therapeutic dose of a narcotic relieves pain and reaction to pain, calms anxiety, and promotes sleep. Common reactions to morphine, heroin, and other morphine-like agents include:

- Reduction in awareness of pain
- Quieting of tension and anxiety
- Decrease in activity
- Promotion of sleep
- Decrease in breathing and pulse rate
- Reduction of hunger and thirst

Some unpleasant reactions to narcotics include sweating, dizziness, nausea, vomiting, and constipation. An overdose of a narcotic results in:

- Reduction in activity and awareness
- Sleep or unconsciousness
- Increasing depression
- Profuse sweating
- Fall in temperature
- Muscle relaxation
- Contraction of pupils to a pinpoint

The continued use of a narcotic produces both psychic and physical dependence.

Stimulants

Stimulants (uppers) are used to increase mental activity and to offset drowsiness and fatigue. The most commonly abused stimulants are

- Amphetamines (Benzedrine — Bennies, Pep Pills)
- Dextroamphetamine (Dexedrine — Dexies)
- Methamphetamine (Methedrine — Meth, Speed, Crystal)
- Methylphenidate (Ritalin)

All of these act similarly and will be described as exemplified by amphetamine. Caffeine is a constituent of coffee, tea, and other beverages. It may produce a very mild psychic dependence, but it does not cause personal or social damage. Cocaine, used medically as a local anesthetic, is a powerful central nervous system stimulant. A new form is called Crack (Rock). It is smoked by inhaling the vapors that are given off as the drug is heated. A therapeutic dose of amphetamine produces the following effects:

- Alertness
- Wakefulness

- Relief from fatigue
- A feeling of well-being

Amphetamines reduce hunger and increase blood pressure. Amphetamine abuse, called a "speed run" involves repeated injections. The daily total sometimes reaches more than 100 times the initial dose. The symptoms include:

- Confusion
- Disorganization
- Compulsive repetition of small, meaningless acts
- Irritability
- Suspiciousness
- Fear
- Hallucinations and delusions, which may become paranoid
- Aggressive and antisocial behavior, which may endanger others

There is little that can be done for the person in this condition except to protect them from injury.

TRANQUILIZERS

The major tranquilizers include Phenothiazines (Chlorpromazine) and Reserpine. The minor tranquilizers are used to calm anxiety and other feelings of stress and excitement without producing sleep. Common examples of minor tranquilizers are

- Meprobamate (Miltown, Equanil)
- Chlordiazepoxide (Librium)
- Ethchlorvynol (Placidyl)
- Diazepam (Valium)

Some tranquilizers are used in treating chronic alcoholism, but in effect, this usage represents substitution of one depressant drug for another. The characteristics of dependence on minor tranquilizers and the related withdrawal symptoms are similar to those produced by barbiturates.

What you have just read about drug and alcohol intake and abuse should convince you that employing a person who intakes alcohol, drugs, or both can cause you many problems. Finding out who they are is a problem. However, your supervisors are in the best position to detect strange and unusual behavior that could be drug and alcohol related.

THE PROBLEM DRINKER

I have heard all sorts of excuses why it is impossible to control drug and alcohol intake, but that is a cop-out. In the first place, take a look at the problem drinker. That person starts to drink the very first thing in the morning, and he is often seen having another drink on his way to work. He has been seen by other employees drinking at work. He has been known to have four to five drinks over a 1 hour lunch period, and that is commonly known by workers as a "liquid lunch". In mid-afternoon, he has been seen taking several belts from a bottle, to tide him over until the end of shift. He is the first one to his car and often fights traffic to buy beer or wine at the nearest store, or he stops at his favorite bar and drinks for an hour or two, and that is just before he gets home; drunk and ready for bed.

Now you might ask, what can be done about this? You cannot do much about his "liquid breakfast" nor his drinking after work, but you can stop the booze from coming into your place by inspection. You can shorten the lunch period to 20 or 30 minutes, instead of an hour. Some companies have an in-house program to help problem drinkers and it helps. You could also make it a condition of employment to be sober at work. Remember this, your supervisors know who drinks.

THE DRUG USER

The drug user is a different problem. You can start to control drug use by subjecting potential new hires to a physical exam. That way you would know if you are about to hire a drug user. For your other employees, you could make it a condition of employment that they be physically examined each year; part of the exam would be to detect drug substances in their urine. Again, your supervisor should observe the behavior of the user; they should look for clues that the users behavior is changing from day-to-day, much the same as the alcoholic that needs a fix to get through the day. I am not advocating that you get rid of your problem by firing the worker, because when you do this, you are giving away your problem to another employer. Think about this, when you hire a new worker, are you taking on someone else's problem?

Many companies have watched productivity take a dive because of drugs. The accident costs have skyrocketed and the number of thefts and resulting losses have dramatically increased. Drug use at work affects eye-hand coordination and judgment which results in damage to equipment and serious injuries. Some indicators of drug use which are self evident:

- Increased absence
- Increased work injuries
- Increased theft
- Late-to-work syndrome
- Decrease in productivity

Your supervisors should be made aware of the above conditions and not ignore the signs of drug use. Everyone, including management, labor unions, and physicians, should tackle this problem. It cannot and should not be ignored nor left to drug screening only. Remember this, it is far better to treat the alcoholic and drug abuser than to subject certain employees to a test. When you decide to test your employees, they should all be tested, and not a random test for a certain few.

SUMMARY

Nearly 40% of workplace deaths and approximately one half of all workplace injuries are directly related to drug or alcohol use. Also, two out of three people entering the work force today have used illegal drugs, according to the National Institute on Drug Abuse (NIDA). Many companies are now utilizing a pre-employment drug screening test when employees show reasonable suspicion of drug use. However, no employer should attempt to initiate a testing program to detect drug use without advice of legal counsel.

This chapter will not make you an expert on drug and alcohol abusers, but it should convince you that the problem is causing a great part of your accidents, injuries, and damage to your property, which in turn is robbing you of your profits.

Credit for a portion of this chapter is given to the American Red Cross.

24

"The secret of success is constancy of purpose."

Disraeli

IS YOUR OFFICE SAFE?

Many employers are so busy with blue collar safety that they ignore white collar safety. All too often, your company's excellent safety record nose dives because of accidents in the office areas.

OSHA does require that office areas must also comply with its regulations. It is a known fact that office people believe that only production and construction workers are exposed to hazards and injuries. Most likely they believe this because office hazards are mild when compared to such hazards as chemicals, electricity, and power machinery.

It is a known fact that hundreds of people die each year in office accidents, and many more are seriously injured. Office safety programs are not only required by OSHA, but also for the following reasons:

- When you try to sell your safety ideas to your managers and supervisors, you expect them to comply; but when they see that your office people do not practice office safety, your top people will believe that you preach, but do not practice safety.
- Your company safety program cannot be totally effective if it covers only a portion of your employees. If your office personnel are considered "exempt", then your production and construction workers will come to believe that complying with your safety rules is a pain-in-the-neck, which will eventually undermine the authority of your supervisors.
- Office injuries are just as painful, severe, and expensive as production and construction injuries. When you do not have a safety program for your office staff, accidents and injuries are far more likely to occur.
- Your office workers are key people and you will not know how important they are to your entire operation until they are suddenly absent, with no immediate replacement available. When this happens, it can throw your payroll, billing, and switchboard operations for a loop.

WHO GETS HURT?

Some years ago, the California State Department of Industrial Relations analyzed reports filed by more than 3,000 California employees. The California study stressed the importance of teaching office workers to look for new hazards and how to correct them. It was also found that there was a substantial increase in the number of injuries in the first year that a company moved into a new office building. The change upset the established routine of the office workers and presented unknown hazards.

New office workers require time to get used to the surroundings. The injury rate among office workers with less than 1 year of service was double that of workers with 1 to 4 years of service, and this latter group's rate was three to four times as high as that among employees with 5 or more years of service.

The same study showed that 10 to 12% of the office help were very young, just graduated from high school, and in their first year of employment. Most of them were females, and not as interested in their job and career as they should have been.

The study also showed that 80% of the disabling injuries occurred to women. The estimated disabling work injury rate for women was about twice as high as that for men.

OFFICE FALLS

Does this title conjure up a scenic cascade of water, visited by thousands of tourists? Well, it is visited by thousands each year, but unfortunately, it is usually a painful experience. Falls make up the largest single type of accident that occur in our nation's offices. Many people believe that because they do not work in an industrial or construction area, they do not need to pay much attention to safety.

According to a study made in California, this mistaken belief allowed office workers to earn the dubious distinction of having 26.6% more falling or slipping accidents than all industrial workers combined.

The following precautionary measures will help reduce the chance of having an office fall:

- Keep file and desk drawers closed when not being used.
- Use aisles, not "shortcuts" between desks.
- Use a stepladder or step stool to reach high places.
- Wipe up wet spots immediately.
- Do not lean back in chairs.
- Carry loads that you can see over.
- Wear shoes with moderate heels.
- Look where you are going.
- Keep wires and cords out of aisles.

LIFTING STRAINS

The California study showed that strains and overexertion occurred almost entirely to men, possibly because of the lesser weights usually lifted by women. Almost three fourths of the strain or exertions occurred while employees were trying to move objects — carrying or moving office machines, supplies, file drawers, office furniture, or other loads. A large number of the injuries resulted from the sudden or awkward movement of the employee and did not involve any object. In other words, reaching, stretching, twisting, bending and straightening up were often associated with these injuries.

STRIKING AGAINST OBJECTS

This caused about 9% of the office injuries in the studies. Two out of three of such injuries were the result of bumping into doors, desks, file cabinets, open drawers, and even other people, while walking. Other causes included hitting open desk drawers or the desk itself, striking open file drawers while bending down or straightening up, and bumping against sharp objects, including office machines and files. Cuts received from handling paper, file drawers, staples, and pins often became infected.

STRUCK BY OBJECTS

This category accounted for about 8% of injuries to office workers in the same studies. Most of these injuries were sustained when the employee was struck by a falling object — a file cabinet that became overbalanced when two or more drawers were open at the same time, file drawers that fell when pulled out too far, office machines and other objects that employees dropped on their feet, or typewriters that fell from a pedestal or rolling stand. Office supplies and equipment sliding from shelves or cabinet tops also caused a few injuries in this class. Employees were also struck by doors being opened blindly from the other side.

CAUGHT IN OR BETWEEN

This final type describes accidents where the employee was caught in or between machinery or equipment. In most instances this involved getting caught in a drawer, door, or window. However, a number of employees were caught in duplicating machines, addressing machines, and fans. Several got their fingers under the knife edge of a paper cutter. Other types of accidents were caused by foreign objects in the eye, spilled hot coffee or other liquids, burns from fire, insect bites, and electric shocks.

A SAFE OFFICE

Your office layouts should be designed for efficiency, convenience, and safety. You should apply the same work flow principles to your offices as to your work areas. Avoid tripping hazards by giving special attention to electric cords and wires. Avoid loose floor mats and runners. Consider using your maintenance workers to move and lift furniture and files, they are trained to lift properly and have the equipment to move heavy objects.

Some reminders regarding basic rules for office personnel are as follows:

- Learn and use proper lifting techniques.
- Avoid overloading top drawers of file cabinets.
- Do not struggle with stuck drawers — get help.
- Keep desk and cabinet drawers closed when not in use.
- Use aisles — do not take short cuts.
- Keep floors free of pencils, paper clips, etc.
- Use stepladders or stools and never chairs to stand on.
- Avoid storing sharp objects in desk drawers.
- Avoid tossing matches, cigarettes, etc., into paper baskets.
- Do not engage in horseplay.

PERSONAL RESPONSIBILITY

Many accidents are caused by one person — do not be that person by observing the following:

- Come to work relaxed.
- Know and follow the office rules.
- Avoid practical jokes.
- Know your strength.
- Watch your step.
- Wear moderate heels.
- Do not take chances.
- Do not be overconfident.
- Do not be impatient.
- Do not show off.
- Report electrical problems.
- Report lighting problems.
- Use handrails.
- Be alert to hazards.
- Stay in shape.

SUMMARY

Your safety program must be a total safety program, and that includes all office personnel. When you provide training for your production and construction workers, you should also instruct your office workers in the hazards of their jobs. When you do this, your officer workers will comply with every safety rule of your company, and your production/construction workers will know that you are indeed serious about safety.

25

OFF-THE-JOB SAFETY

"Chance fights ever on the side of the prudent."

Euripides

RECREATIONAL SAFETY

Every year over 20 million Americans are injured in recreational activities. Taking part in leisure activities can be fun, but your employees can get hurt and bring their injuries to your workplace. How can you prevent this from happening? In the first place, accept the fact that your employees will take part in off-the-job recreational activities, but provide them with the information they need before they decide to engage in leisure sports.

Your employees should consider four things before they take part in any activity:

- Knowing their limits — not pushing the body beyond its capability.
- Awareness of the hazards — every sport or activity has a potential for injury.
- Getting in shape — keeping weight down with proper exercise.
- Medical check — getting an okay from their physician.

Some people start recreational activities with a bang, and without a consideration to a long period of inactivity, especially during the winter months. People who do this are courting injuries. Consider some common injuries associated with recreational activities:

- Blows when colliding with a fixed object or with another person.
- Slips and trips because of quick turns or poor footwear.
- Sprains and strains to ankles, legs, neck, back, and wrists.
- Sore muscles, blisters, and aches.

HOME ACCIDENTS

Home accidents kill about 30,000 people and injure more than 4 million each year. Nearly all of these deaths and injuries can be avoided as follows:

- Provide ample lighting.
- Repair loose floor coverings right away.
- Avoid clutter and practice good housekeeping.
- Wipe all spilled water, oil, grease, paint, etc.
- Avoid placing electrical cords in walkways.
- Do not create obstacles with furniture.
- Install handrails for all stairs.
- Secure all scatter rugs.
- Install shower and tub hand grabs.
- Use nonskid mats in bathroom.
- Get help when moving furniture.
- Keep medicines out of reach of children.
- Store household chemicals out of reach of children.
- Do not plug more than one appliance into same outlet.
- Repair defective electrical appliances right away.
- Do not allow rubbish to accumulate.
- Check chimney, pipes, flues, fireplace, furnace and wood stoves.
- Do not smoke in bed and empty all ashtrays after use.
- Place stickers on glass doors.
- Do not store paint, gasoline, etc., in unvented spaces.
- Store pesticides in locked lockers.
- Use a wooden ladder.
- Never pull a power mower toward you.
- Check for poisonous plants and insects.
- Handle and store food properly in warm weather.
- Avoid sunburn, heat exhaustion, and sunstroke.
- Stay in house during lightning storms.
- Keep fire extinguishers in good condition.

CUTTING WOOD

A chain saw is made for serious work, therefore, it should always be used seriously. A chain saw is considered to be a hazardous tool because in a single second, it can cause a very serious injury or death. Before you decide to cut wood, make sure that you follow these basic rules:

- Wear a hard hat or a widow maker (falling branch) will get you.
- Wear safety glasses or face shield.
- Wear hearing protection.
- Wear nonslip heavy-duty gloves.
- Wear close fitting clothing.
- Wear heavy work boots with safety toes.

You should be aware of weather conditions, especially on wet, snowy, or cold days. Cutting on windy days requires extra caution because the wind can cause trees to fall in the wrong direction — toward you. The wind can also blow down dead branches on top of you.

Plan your job before you start (you should read the owner's manual before you start cutting) and never cut alone. You should never cut trees or wood when you are tired, and you should always avoid alcohol or medication that could affect your reaction to danger. You should be very careful when fueling your saw and never refuel a running or hot saw because of the danger of fire or explosion. Check your saw chain to make sure it fits snugly and does not bind to the bar.

You should never cut with one hand. Use two hands and place both feet firmly to the ground when cutting. Never stand directly behind the saw, but instead to one side. You should never walk with a running saw. Check your owner's manual for proper notch-cut and back-cut methods. When the tree is down, do not cut limbs that are propping the tree or log, but instead, roll the log to place the limbs on top. You can avoid back strains by using hooks, pry bars, or a P.V. to roll logs too heavy to lift. Do not wait until next time to sharpen your saw because you might forget — a dull saw is a dangerous saw. Think ahead each time you cut a tree. The life you save could very well be your own.

DEFENSIVE DRIVING

Auto accidents do not always happen to the next guy. Someday it could happen to you. It could be someone else's fault, but most two-car accidents can be avoided by using greater caution and the following safe principles:

- Drive defensively — anticipate a coming accident.
- Be courteous — do not lose your temper.
- Use skill — keep at a safe distance.
- Be under control — do not drink alcohol, be alert, and stop if sleepy.
- Drive a safe car — check your car often.

Do not assume that because your light is green that the other driver will stop at a red light. The same applies at stop signs. Always drive according to weather conditions and do not try to pass long lines of cars.

Dim your lights even when the opposing driver does not. Give pedestrians a break; after all, they cannot move as fast as you can. Always use your signals when changing lanes. Do not tailgate the driver ahead of you. Do not pass on the right, and do not be afraid to use your horn.

Keep your eyes moving and do not be distracted. Keep the whole traffic situation in view at all times; ahead, back, and to the sides. Be prepared to reduce speed in heavy traffic by not using your speed control device. Always make sure that you are seen by the cars around you.

Do not drink alcohol, you need to be in self-control with a clear head. Alcohol mixes well with many things, but never with driving. Do not daydream, it can take your mind off the road. Do not be in a hurry; remember that haste makes waste. When you become tired or sleepy, pull off to the side and walk a bit.

Your car should be in tip-top shape. You need good brakes, working windshield wipers, proper tire pressure, working lights and horn, your back windows should be clear, and your exhaust system in top shape. At the first sign of mechanical problems, get your car off the highway and signal for help by raising the hood or lighting a flare at night. If you have to change a tire, pull well off the road and warn oncoming traffic with a reflector or flares. At night leave your parking lights on.

Remember this! Every year accidents kill at least 45,000 people and injure almost 2 million people. The annual cost is estimated to be about $20 billion. Last, prepare your car for winter driving, it is a whole different ball game.

PEDESTRIAN SAFETY

Most of the time a pedestrian has the right of way. The reason being, that the automobile can move faster out of harm's way than a pedestrian. However, pedestrians do have certain responsibilities to safeguard themselves. The pedestrian should be courteous and should observe traffic laws, stop signs, and lighted signals, but you should also have some consideration for drivers. Never forget that it is the pedestrian who is most likely to be hurt or killed in any traffic accident. A pedestrian should move quickly through crossing and not jaywalk. As a pedestrian, you must be willing to yield when push comes to shove. It will not matter if you are dead right or that the driver is dead wrong. In either case, you will be dead.

Do not run on sidewalks or at crossing just to beat other pedestrians. You may knock them over and get hurt yourself in the process. Allow plenty of time to walk to your destination safely — it is a better way to start your day. Always cross at the crossings and never diagonally through the crossing. If you do not, you will be pushing up daisies before you have to. Always wear white or bright clothing at night, and always walk on the left facing oncoming traffic, unless of course, there are sidewalks on both sides of the street or road.

In bad weather, always dress accordingly. During winter, wear good footwear and avoid leather soles on slippery or icy surfaces. Do not allow your children to play in the streets — you might regret it.

BICYCLE SAFETY

Bike riding is fun, but it can be very dangerous. Always remember that not unlike the pedestrian, you cannot move as fast as a motor vehicle. Almost half a million bike-related injuries occur each year, and nearly 1,000 cyclists are killed each year. It is very hard to see except straight ahead on a bike. Some common causes of bike accidents are

- Loss of control when going over uneven surfaces.
- Mechanical problems, such as brake failure.
- Clothing caught in chain, gears, and wheels.
- Chance-taking by cutting across traffic and through lights.

Just like with a car, you should run a safety check of your bike before you use it. You can tell if something is wrong by the sound it makes. Make sure that your tires are properly inflated and that your bike is well lubricated. You are required to have reflectors on your bike. Your bike's headlight should be visible from at least 500 feet and your rear reflector should be visible from at least 300 feet. You should dress with bright colored clothing to help you be more visible at night. You should know the capabilities and limitations of your bike. Also, do not ride your bike barefooted; it is dangerous.

Bicyclists are required to obey the same rules as motorists when approaching stop signs, traffic signals, and when making right and left turns. Always keep to the right of the road and travel in single file with the traffic. Do not forget to yield to cars because they are faster, bigger, and harder to maneuver. Also, it is always a good idea to stop when being chased by a dog, and be careful of kids.

SUMMARY

There are many many other off-the-job safety subjects that would require enough writing to fill a book. Off-the-job safety affects on-the-job safety. You cannot be safe with one and not with the other. When your safety and health program includes off-the-job safety, you are helping your program. Your workers will also know that you care. It is not difficult to show that you care. You can purchase little booklets on off-the-job safety and pass these out with paychecks, or you can place an off-the-job safety article on your bulletin boards. You can also print your articles in your company newspaper. Whatever you do, it will save you money when your employees subscribe to safety practices 24-hours a day.

26 PREVENTABLE INJURIES

"It takes less time to do a thing right than it does to explain why you did it wrong."

Longfellow

BACK INJURY

Approximately 400,000 workers suffer disabling back injuries each year and receive estimated compensation and medical payments in excess of 1 billion dollars. It has been said that, "A man's best friends are his wife, his back, and his dog." The back, however, has a reputation of not being as faithful as the other two friends.

Overexertion in the workplace accounts for a large number of disabling injuries. Most of these injuries involve the act of manually handling materials. The National Safety Council reports that at least 25% of all injuries, accounting annually for 12 million lost work days and over 1 billion dollars in compensation costs, can be attributed to such mishaps.

Admittedly, human suffering and the high economic burdens are greatest in the industries that require manual handling of objects. Back injuries can cause you and your workers pain, lost time, expense, inconvenience, and disability. Most back injuries are caused by improper lifting practices and overload. The proper lifting practice is

- Stand close to the object with feet firmly placed.
- Squat down and straddle the load, keep back straight, and knees bent.
- Grasp object firmly with a sure grip.
- Breathe in and fill your lungs.
- Lift with your legs and slowly straighten them.
- Hold the object close to your body.
- Make sure your path is unobstructed.
- Get help if the object is too heavy.

Keep in mind that your workers' physical condition plays an important part in back injury prevention. The following poor physical conditions make your workers prone to injury:

- Poor posture — keep head up and stomach in.
- Lack of exercise — get in shape.
- Excess weight — watch your weight.

HAND AND FINGER INJURY

The National Safety Council has said that for table saw operators who have been on the job for 15 years, 9 of every 10 operators will have lost fingers. An unguarded machine capable of cutting, bending, or shaping steel can turn on you with a vengeance, and the admonition, "I warned you" is of small comfort to an operator left with no fingers. A broken and whipping band saw blade and the curling and whipping metal shaving from a metal lathe will show no mercy to the hand or fingers that get in the way.

The object here is not to frighten your workers or trainees with lurid details of what can happen when a machine is unguarded, but instead to instruct each worker in the use of the machine guarding so that they may operate a machine safely and efficiently. Other injuries to hands and fingers are caused by hot objects, cold objects, electrical shocks, and by absorption of chemicals. These injuries can be prevented by the use of gloves. There is a wide assortment of gloves, hand pads, and wristlets for protection from various hazardous situations. The protective devices should be selected to fit the job. For example, some gloves are designed to protect against specific chemical hazards. Your workers may need to use gloves which have been tested and provide insulation from burns and cuts such as wire mesh, leather, and canvas.

Your employees should become acquainted with the limitations of the gloves used. Certain occupations call for special protection. For example, electricians need special protection from shocks and burns. Rubber is considered the best material for insulating gloves.

Gloves should be long enough to come well above the wrist, leaving no gaps between the glove and the shirt sleeve. Soft cotton gloves should not be worn around rotating machinery such as drill presses, as the drill could grab the gloves and amputate the fingers.

The machines that cut with shearing motion, those that have rollers or gears that grab, and those with two solid objects coming together to crush or stamp, should be guarded or equipped with two hand controls.

Hand and finger injuries can be prevented by installing machine guards and providing hand protection. However, it will not do any good if their use is not enforced.

FALLING INJURIES

Each year over 13 million workers are injured in falls, and over 14,000 die as a result of the fall. Many workers fall from improperly placed ladders which should be placed at the bottom 1 foot from the building for every 4 feet of height. A ladder reaching 20 feet would be placed away from the building 5 feet at its footing. The ladder should be checked for broken, missing, or loose rungs. It should be free from oil, grease, tar, or paint. The ropes, pulleys, and foot pads should also be checked. Do not overextend extension ladders — keep an overlap of at least three rungs at the mid-section. When climbing, keep both hands on the side rails, wear good rubber soles, and make sure they are not wet, greasy, or oily.

No more than one person should be on the ladder at any time. Do not carry anything in your hand, instead, use a rope to haul material up to you or onto the roof. Never stand on the top two rungs of a ladder and always extend the ladder at least 3 feet above the roof to provide a secure hand-hold when stepping off the ladder to the roof. Never stand on the top-flat of a stepladder.

Safety belts, lanyard, and lifeline are required by OSHA when working at a height of 6 feet or more. Another fall prevention requirement when working at a height of 25 feet or over without a safety belt requires that a safety net be installed. Guard rails are required on scaffolds located 10 feet above the ground. Floor and wall openings must be guarded by a railing on open sided floors.

Platforms and runways must also be guarded by a railing when a drop is more than 6 feet. OSHA has other requirements for fall prevention from low pitch roofs when the height from roof-to-ground exceeds 16 feet, and they are as follows:

- Motion-stopping safety systems (mechanical system)
- Warning lines 6 feet from edge of roof
- Safety monitoring system (competent person)
- Safety belt system with catenary line or anchorage

Falling accidents can be prevented only if you insist on the above requirements.

SLIPPING AND TRIPPING INJURIES

You can avoid slipping and tripping accidents by keeping an eye out for hazardous walking conditions. Most slipping injuries are caused by oily and greasy work floors. Your workers must be trained to wipe any spills in their immediate work areas. Oil and grease can be cleaned with rags and detergent. Do not let grease and oil accumulate on shop floors; spread sawdust or oil-absorbent materials such as speedy dry on oil and grease right away.

When water is spilled, do not wait for the water to dry by itself. It only takes a second for a serious accident to happen — wipe it up right away. Sometimes slipping accidents happen in the winter when water turns to ice. When this happens, spread sand on icy walkways and caution your employees to walk slowly. It is also a good idea to walk with your feet pointing out to the sides for greater stability.

Tripping accidents are caused by poorly installed stair-steps or stairs without railings. The risers of each step should be not less than 6.5 inches in order to prevent tripping. Walking and working surfaces should be even and not polished. Carpets should be tightly installed in office spaces. Loose flooring should be fixed or a caution sign should be placed in the immediate vicinity to warn employees of the hazard.

Electrical cords or conduit should never be placed in the walkways of workplaces. Materials should never be stored in aisle ways and stairways. Untidy floors can cause an injury. Even a small object like a nail or bolt can trip a person and cause an injury. Do not skimp on lighting; unlighted work areas can cause one of your workers to slip, trip, and fall. Tripping injuries can be prevented when supervisors take time each day to look over their work-sites.

ELECTRIC SHOCK INJURIES

The Bureau of Labor Statistics reports that hundreds of workers die each year from electrocution and many thousands more are seriously injured by electrical shocks. What makes this more tragic is that, for the most part, these accidents and fatalities could have been prevented. Electricity travels in closed circuits, and its normal route is through a conductor. Shock occurs when the body becomes a part of the electrical circuit. The current must enter one part of the body and leave through another part of the body. Shock normally occurs in one of three ways. The person must come in contact with:

- Both wires of the electrical circuit.
- One wire of an energized circuit and the ground.
- A metallic part that has become "hot" by being in contact with an energized wire, while the person is also in contact with the ground.

The metal parts of electrical tools and machines may become "hot" if there is a break in the insulation of the tool or machine wiring. The worker using these tools and machines is made less vulnerable to electrical shock when a low resistance path from the metallic case of the tool or machine to the ground is established. This is done through the use of an equipment grounding conductor — a low resistance wire that causes the unwanted current to pass directly to the ground, rather than through the body of the person in contact with the tool or machine. If the equipment grounding conductor has been properly installed, it has a low resistance to ground, and the worker is therefore protected.

Electrical accidents appear to be caused by a combination of four possible factors:

- Unsafe equipment and/or installation
- Workplaces made unsafe by the environment
- Unsafe work practices by employees
- Lack of maintenance of electrical equipment

There are various ways of protecting your people from hazards caused by electricity. These include:

- Insulation with a high resistance to electrical current.
- Guarding or isolating live electrical parts from people.
- Grounding the equipment to earth or ground plane.
- Safe work practices by de-energizing electrical equipment before inspecting or making repairs.

EYE INJURIES

Of all the injuries that can happen to your workers, eye injuries are the most debilitating. Eye accidents occur at work from countless sources. It is a major problem to decide what eye protective measures should be adopted and when and where these should be worn. In some situations your workers should wear eye protection all the time. At other times, eye protection should be required only on certain jobs.

In any event, our eyes are our most valuable possession. The softness of the eye provides the easiest entry into the brain. Over 100,000 eye injuries occur each year. Because our eyes are the windows of our mind, no one should be allowed to take chances. Eye protection must be provided and used. The prevention of eye injuries is based on these major fundamentals:

- Establishment of safe work practices, whereby proper eye protection will prevent an eye injury
- Installation of devices to control the hazards at their sources
- Providing employees with eye protection when exposed to eye injury hazards
- Enforcing the requirement that eye protection will be worn when exposed to eye injury hazards

It is also important to provide eye protection to passers by and visitors. This may be accomplished with visitor-type eye protection and for passerby protection, you should consider the use of portable screens.

The most common causes of eye injuries are as follows:

- Flying chips, fragments from grinding operations, nails, splinters of wood, sawdust, plaster, and concrete chips
- Small flying objects from sandblasting operations and air-borne dusts
- Splashes of molten metal, soldering, casting, pouring, and brazing
- Splashes when handling liquids, acids, and corrosives
- Hazardous dusts, gases, and vapors which affect the eye directly
- Radiant energy, welding, brazing, etc.

Think about this, 2,000 eyes are lost in the United States each year through accidental injuries which cost our industries about $37 million in medical expenses and compensation.

HEAD INJURIES

The Bureau of Labor Statistics reports each year that most workers who receive head injuries were not wearing head protection and it was not required by their employers. The typical injury was caused by a falling metal object weighing at least 8 lb. In almost half the accidents, employees knew of no actions being taken by employers to prevent the injuries from occurring again. Hard hats, when properly selected, will greatly reduce injury from falling objects, as well as exposure to electrical burns.

Another form of protective head gear is the bump cap, a thin shelled, lightweight plastic cap. It is very effective when your employees are working in tight quarters or confining spaces. For the general industry, OSHA, Part 1910.135, requires that hard hats will be worn when workers are exposed to falling and flying objects and limited electric shocks and burns. For the construction industry, OSHA, Part 1926.100, requires that hard hats will be worn when workers are exposed to falling and flying objects or electrical shock and burns.

SUMMARY

When you have taken action to reduce all the injuries and deaths associated with the above hazards, you will be on your way to reducing your accidents and increasing your profits. Worldwide, successful company managers have committed themselves to reducing injuries by providing adequate personal protective equipment for their workers. They have also instituted a program for enforcement of their safety rules and policy. Believe me, it works.

27 TRAGIC ACCIDENTS

"Courage consists not in blindly overlooking danger, but in seeing it, and conquering it."

Richter

THE TRAGEDY OF A DEATH

Avoiding tragic accidents is not only a requirement of law, but also a humanitarian right. I have seen what I believe to be the six most tragic types of deaths, and as long as I live, I will never forget any of them. I have seen people:

- burned to death
- who fell to their death
- suffocated to death
- bled to death
- drown to death
- shocked to death

When you have witnessed one of the above tragic deaths, you will remember that event forever. It is a tragedy that is not easily forgotten, and it will disturb you every day of your life. It is so tragic that your employees will never let you forget that it was your fault. It does not matter that it could have been the worker's fault, you will still get blamed because they feel that you could have prevented the event in the first place.

BURNED TO DEATH

Burning to death is a horrible way to end a life, yet it can be prevented. Chapter 9, Fire Prevention, is a very important element of your safety program. It can only be effective when you put in place a program designed to prevent fires from starting. Most industrial fires are caused by poor housekeeping and/ or the careless or uncontrolled use of flammable liquids. Fires are also started by improper electrical hookups which may go undetected unless you institute a planned program of safety inspections. Flammables must be stored in approved lockers. As a matter of fact, it is also a requirement of the Hazard Communication Act, which is now a law that you are required to comply with. Allowing your welders or burners to work alone without a fire watch is chance-taking and could bring on a tragic occurrence. The fire watch has many duties to perform:

- Assist the welder or burner
- Stand by with a fire extinguisher
- Put out a small fire
- Call for assistance
- Rescue the welder
- Sound the fire alarm

Allowing your employees to smoke in non-smoking areas is inviting a tragedy to happen. You must make certain that your supervisors are enforcing your non-smoking policy. In addition to your regular safety inspections, your maintenance department must be on the lookout for fire hazards. They are in the best position to detect unauthorized electrical hookups, poor housekeeping, and improper use of flammables. Fire drills, along with the above policy, should prevent a person from being burned to death.

FELL TO DEATH

Falling to one's death is also a terrible way to die, and it is doubly unfortunate because it could have been prevented. Whether the work is performed on a roof or the side of a building, the danger of death from a fall is always present. A worker can also fall from a staging, scaffold, crane, aerial basket, elevator shaft, etc. How can falling accidents be prevented? In the first place, do not send workers aloft when you know they fear heights — that is chance-taking.

Employees working on roofs with eaves 16 feet or more above the ground, and with a pitch of 4 inches for each foot or greater, must be protected by the use of a catch platform or safety belt and lanyard. The catch platform must be equipped with a guardrail, midrail, and toeboards. The catch platform must be fully planked and should extend at least 2 feet beyond the eaves. If a safety belt and lanyard are used, the worker must be instructed in their use. Most staging and scaffold accidents are caused by overload.

Stagings and scaffolds were not meant or designed to become storage platforms. The workers should only take up to stagings and scaffolds immediately needed material. It is also a requirement to wear a safety belt and lanyard when working from an aerial basket (cherry picker). The lanyard must be attached to the basket in order to prevent the worker from overreaching and falling.

Floor openings of buildings must be guarded by a standard railing, midrail and toeboards, or the installation of a cover over the opening. All wall openings must be guarded by railings and midrail when the drop is over 4 feet. When you have instituted a policy for prevention of falls, and it is enforced by your supervisors, you will have taken the steps necessary to prevent a falling death.

SUFFOCATED TO DEATH

Suffocating to death is a dreadful way of dying. It can happen several ways. I will only mention two ways that apply to industry and construction. Working in confined spaces such as tanks, underground tunnels, and storage bins can be lethal. Needless deaths occur each year because workers are overcome by unseen hazards after entering confined spaces to perform routine work. For example, a worker entered a storage tank and later on was found at the bottom of the ladder inside the tank. The culprit was not the tank nor the fact that the tank had not been cleaned or ventilated, but instead the fault was failure to test the space prior to entering and working in that confined space. Instituting and enforcing procedures for entering and working in confined spaces can help ensure your workers safety.

Prior to entry into any confined spaces, your workers should be trained to know and ascertain the following:

- Does an oxygen deficiency exist?
- Are toxic or flammable gases and vapors present?
- Is the space to be entered clean and ventilated?
- Are proper type of respirators available?
- Will constant communication be maintained?
- Will safety lifelines be installed?
- Are "No Smoking" or "No Burning" signs available?

The workers should be trained to the requirements of OSHA prior to entering any confined spaces.

Trench and excavation cave-ins also cause many workers to suffocate each year, and they too are preventable. Too many contractors and the workers they employ fail to realize the danger of working in unprotected or poorly protected excavations. With little or no warning, an unsupported or improperly shored and sloped trench or excavation wall can collapse, trapping the worker(s) below in seconds. Inadequate shoring in an attempt to cut costs or save time, misjudgment of soil conditions, defective shoring materials, failure to evaluate changing weather conditions or heavy loads in the area — these are among the common causes of trench and excavation cave-ins.

According to OSHA, a trench is a narrow excavation in which the depth is greater than the width, although the width is not greater than 15 feet (an excavation is any manmade cavity or depression in the earth's surface). OSHA requires that all excavations over 5-feet deep be sloped, shored, sheeted, braced, or otherwise supported. When soil conditions are unstable, excavations shallower than 5 feet must also be sloped, supported, or shored.

Remember this, OSHA regulations (Subpart P-Part 1926.650, leaves no room for risk-taking. Being buried alive is a terrifying experience to the person who lives through it — and most do not.

BLED TO DEATH

Bleeding to death is a frightening way to die. Loss of more than a quart of blood is a threat to a person's survival. Hemorrhaging from the aorta (the largest blood vessel of the body), or from combined external and internal injuries may be so rapid and extensive that the victim dies almost immediately.

In any severe bleeding injuries, the injured person is not able to get first aid within 3 to 5 minutes, therefore, the victim bleeds his life away.

Most common causes of wounds are motor vehicle accidents, falls, and the mishandling of sharp objects, tools, machinery, and weapons. Some examples of causes are wood cut-off saws and powered shears.

Others are caused by punctures from automatic stud and nailing guns, while others are caused by machines that can grab and tear the limbs and flesh. The list of tools, equipment, and machinery that can cause a deep wound and extensive and rapid bleeding is too long to list. Most of the injuries happen to the untrained and/or new workers. Make sure your workers receive training before they are allowed to use tools, machinery, and equipment. It is an awful blow to see a person bleed to death.

DROWN TO DEATH

A drowning death is so pitiful to witness, you never forget it. Drowning is a type of asphyxia related to either the aspiration of fluids or obstruction of the airways caused by spasm of the larynx while the victim is in water. Drowning is a major cause of accidental death in the United States. Drowning accidents can and do happen to construction and shipbuilding workers.

A drowning person may be seen either struggling in the water and making ineffectual movements, floating face down on the surface of the water, or lying motionless under water. Many persons sink very quickly as they lose buoyancy by swallowing water and by aspirating it into the lungs where it replaces air. The victim sinks beneath the surface and begins to lose consciousness from asphyxia. Most construction and shipbuilding drowning accidents occur when workers are required to work over water, near water, on ship decks, alongside piers, bridge supports, etc., without protection.

OSHA, in Subpart E, Part 1926.105, requires that safety nets must be provided when workplaces are more than 25 feet above the surface of water or where safety belts, lanyards, and safety lines are impractical. OSHA, Part

1926.106, requires that employees working over or near water, where danger of drowning exists, shall be provided with U.S. Coast Guard-approved life jackets or buoyant work vests. OSHA further states, prior to and after each use, the buoyant work vests or life preservers must be inspected for defects which would alter their strength or buoyancy. Defective units shall not be used. Also ring buoys with at least 90 feet of line shall be provided and readily available for emergency rescue operations. Distance between ring buoys shall not exceed 200 feet.

Last, at least one lifesaving skiff (boat) shall be immediately available at locations where employees are working over or adjacent to water and the persons using the boat will be wearing life jackets or buoyant work vests.

There is no reason to lose a worker to a drowning accident. Every situation is different, and OSHA's rules are minimum requirements, therefore, it is possible that some work situations would add other protective requirements and escape plans in cases of flooding.

SHOCKED TO DEATH

An electrocution is a dreadful mishap. Electrical energy is a good and faithful servant, but when not used properly, it can turn on you. The Bureau of Labor reports that each year, thousands of workers die of electrical shock. Electricity travels in closed circuits and its normal route is through a conductor. Shock occurs when the body becomes a part of the electrical circuit. The current must enter the body at one point and leave at another.

The person must come in contact with both wires of the electrical circuit, one wire of an energized circuit and the ground, or a metallic part that has become "hot" by being in contact with an energized wire, while the person is also in contact with the ground.

When there is a break in the insulation of a tool or machine wiring, the metal parts and machines may become "hot". OSHA requires contractors to use ground fault circuit interrupters (GFCI) at construction sites where there is likelihood that a worker could receive an electrical shock because of improper wiring, insulation breakdown, and when working in wet locations.

The GFCI is a fast acting circuit breaker which senses small imbalance in the circuit caused by current leaking to ground and, in a fraction of a second, shuts off the electricity.

There are many reasons why you should not expose your workers to electrical danger, but one that sticks in my mind is when I saw the burns and charring on a victim's body after he came in contact with electrical energy. The person was a licensed electrician for over 20 years, but on that day, he took a shortcut and ignored the rules and training. More information regarding electrical safety can be found in OSHA, Subpart S, for industry, and OSHA, Subpart K for construction.

SUMMARY

The six types of deadly accidents should convince anyone that they should and must be avoided. Besides being tragic, they are costly and demoralizing to your work force. No employer should take a chance with faulty equipment, nor should they allow their workers to take shortcuts at the expense of safety. Training your employees is very important, even retraining for your older workers. A good way to get the point across that you expect your workers to obey safety rules is when you require your supervisors to give safety talks and stand-up safety meetings, or at construction sites when the safety meeting is called "tail-gate talks." The tragedy of a work-related death is something you will want to avoid because the expense outnumbers any safety improvements you could have made to avoid the death in the first place.

28 SELF-EVALUATION

"Unless a man undertakes more that he can possibly do, he will never do all that he can do."

Henry Drummond

INTRODUCTION

Your Occupational Safety and Health Program is an essential activity. Its goals are to provide a work and physical environment for your employees that is as free as possible from potential hazards. Good management and work practices, as well as legal requirements, make safety and health priority concerns for you and your employees.

LEGAL REQUIREMENTS AND SELF-EVALUATIONS

Self-evaluation, with subsequent corrective action, should provide satisfactory assurance that your establishment has identified and addressed the problems that would be cited by a Federal or State inspector. Most companies recognize the need to comply with the law, but many find it difficult to interpret the law and define their responsibilities.

This process will help you to understand not only your legal responsibilities, but will assist you to appreciate the benefits of going beyond the minimum requirements of the law.

OVERVIEW

The size and scope of your safety and health program will depend upon the size of your workplace, the number of employees, your geographic location in relation to emergency care facilities and medical clinics, and the number and extent of potential hazards inherent in your operations. For optimum effectiveness, regardless of size and number of employees, your program must have certain major elements:

- A comprehensive safety and health policy
- A written set of basic objectives
- A functional organizational chart and responsibilities
- A commitment from management
- A plan for shared responsibilities
- A system with identifiable benchmarks for reporting progress

SETTING GOALS AND OBJECTIVES

The goals of your safety and health program should reflect the priorities of your workplace and should serve as the underlying principles for more specific policies and procedures. This list of goals might address such priorities as:

- Assuring employees of safe and healthful working conditions.
- Proper placement of employees according to their abilities.
- Providing medical care and rehabilitation for injured employees.
- Encouraging employees to maintain good health.

These principles must be clearly stated. They need not be lengthy or complex. In setting program objectives, you should seek employee input. A questionnaire might be developed and circulated to all employees to obtain their ideas. This will ensure employee involvement in establishing the program itself, and will provide you with first-hand knowledge of problems and situations that should receive attention. The feedback you receive from this survey should be analyzed to identify common concerns and interrelated problem areas. After this is done, you can rank these objectives and develop time frames for their attainment.

SPECIFIC PROGRAM ELEMENTS

Depending on the size of your establishment and the nature of your operations, your program will need to include elements of first aid, safety, industrial hygiene, and occupational health. Although large workplaces may have professionals in these disciplines on their staff, it is unlikely that smaller companies would have the need or resources to hire full-time professionals in these areas. In these instances, the services of insurance companies, local health departments, industrial clinics, governmental agencies, or independent consultants could be used for specialized assistance. With training and experience, the staff you now have could perform some environmental evaluation techniques such as noise monitoring or air sampling.

SAFETY

A safety program is not imposed on company organization, but must be built into every process or product design. The critical elements in any successful programs are

- Employee involvement.
- Management leadership and support.
- Assignment of responsibility.
- Maintenance of safe working conditions.
- Establishment of safety training.
- An accident record and investigation system.
- Medical and first aid program.
- Acceptance of personal responsibility by all employees.

Managerial interest in safety must be sincere and constantly visible; without this, even the best policies will be meaningless. No matter how small your organization, any attempts you make to stop accidents without a definite guiding policy will fail and you will find yourself continuously "fighting fires". Your safety policy should contain a few essential assumptions:

- Employee awareness and involvement are keys to program success.
- Safety is paramount and will take precedence over any shortcuts.
- Every attempt will be made to reduce the possibility of an accident.
- Your company intends to comply with all safety and health laws.

Behind these statements must lie a total safety program including:

- Development and application of safety standards for equipment, work methods, and products.
- Safety inspection to identify potential hazards, both in production and products.
- Accident investigation to determine future preventive action.
- Accident records and accident cause analysis to determine accident trends.
- Education and training in safety principles.
- Personal protective equipment to provide injury protection.
- Safety publicity to step up program interest and participation.
- Off-the-job accident prevention.

HEALTH

Regardless of the size of your establishment, a good occupational health program should maintain the health of your work force, prevent or control diseases and accidents, and prevent and reduce disabilities and the resulting lost time.

Specifically, your program should provide the following:

- Disease prevention, industrial hygiene, and safety.
- Early detection and treatment of occupational illness.
- Emergency medical care.
- Rehabilitation of disabled workers.
- Placement of workers in accordance with their physical and mental abilities.
- Maintenance of workers' physical fitness.
- Medical records with complete information on employees' physical conditions.

INDUSTRIAL HYGIENE

The science of industrial hygiene deals with the identification, evaluation, and control of those physical, chemical, and biological agents and psychological stresses in the workplace that may affect workers' health. These agents include such common workplace hazards as:

- Toxic materials that may be breathed, swallowed or absorbed into the body as fumes, vapors, dusts, or mist.
- Noise and vibrations.
- Ionizing and nonionizing radiation.
- Extremes of temperature and humidity.
- Light.
- Bacteria and fungus.
- Viruses.
- Plant and insect pests.
- Shift work stress.

An industrial hygiene program involves the use of sampling equipment, measurement instrumentation, and laboratory analysis to assess your occupational environment.

Most smaller companies cannot afford to maintain all the equipment and facilities they need; they must depend upon outside services and rental for much of their program. An industrial hygiene program should provide:

- Plant inspection and identification of health problems.
- Measurement and evaluation of these potential problems.
- Information to both management and employees.
- Advice to management for effective engineering and administrative controls.
- Coordination of corrective action and evaluation.
- Application of ergonomics (human factors).

FIRST AID, PHYSICIANS AND NURSING SERVICES

Every company, no matter how small, should have one person on each shift who is trained in first aid. In small companies this person will undoubtedly have regular duties not related to first aid, whereas in a large establishment, a full-time first aid person might be employed. In selecting first aid people, you should look for employees who are people-oriented and who have demonstrated some leadership qualities. The employee must have formal training in first aid. You can obtain more information through your local Chapter of the American Red Cross.

Every company not directly employing a full or part-time physician must have available the services of a consulting physician. This might be a doctor from a nearby hospital with which you have emergency care arrangements, or a private practitioner in the community oriented to occupational medicine. The consulting physician must approve all first aid supplies and be available to give medical directions to nurses employed by or serving your establishment.

Many small companies employ a part-time nurse directly or by arrangement with a local nursing service that can send a nurse to the plant on a regular basis, or visit the plant on request. It is important to always get the same nurse so that they can maintain familiarity with your working conditions. Larger companies, or those with high levels of potential hazards, should employ a full-time nurse. For the maintenance of complete health service, it is recommended that there be 1 nurse for up to 300 employees, 2 or more nurses for up to 600 employees, and so on. The nurse should be a graduate of an accredited school of nursing and licensed to practice in the state where he or she is employed.

RECORD-KEEPING

Accurate injury and illness records are invaluable in providing information for your safety and health program. Since injury and illness recording is already a requirement under law (i.e., OSHA Forms 101 and 200), you should use this information to your advantage to:

- Reveal which operations are most hazardous
- Determine weaknesses in your safety and health program
- Judge the effectiveness of your program by comparing it with past records
- Aid in accident analysis and investigation
- Identify the causes of occupational diseases
- Satisfy legal and insurance requirements

INSPECTIONS

These inspections should be both formal and informal and include both employee and supervisors. Together they should develop check lists, schedules, and guidelines. If you use employees most familiar with the processes, machines, conditions, and practices, this will promote employee involvement and help them recognize their responsibilities at their work-sites. If your facility is kept as clean as possible, this will improve employee morale and help in the inspection process by not covering up potential hazards. Although you will need to develop your own check list, the following list of basic work hazards can be used as a guide:

- Pinch points
- Catch points
- Shear points
- Squeeze points
- Flying objects
- Falling objects
- "Run-in" points
- Electricity
- Gases
- Heavy objects
- Chemicals and flammables
- Hot and cold objects
- Radiation
- Sharp and pointed objects
- Slippery surfaces

BUILDING SAFETY INTO YOUR OPERATIONS

Your company policy should specify that employee protection must be designed and built into the job:

- Purchase only equipment with safety features
- Consider the study of ergonomics for employee comfort
- Institute a workplace policy regarding alcoholism and drug abuse
- Set reasonable lifting limits for your employees
- Develop training programs
- Install local and general ventilation
- Install proper lighting
- Maintain good walking surfaces
- Maintain safe stairways and railings
- Provide adequate exits
- Maintain a safe electrical system
- Provide sanitation facilities for your employees
- Provide personal protective equipment
- Install machine guards
- Provide eye wash and deluge showers
- Maintain a hazard communication program
- Maintain fire prevention
- Conduct fire drills

Credit for some information is given to the National Institute for Occupational Safety and Health (NIOSH).

SUMMARY

Any program or activity in which you invest time and resources on a continual basis should prove its worth. A well structured safety and health program should lend itself to objective performance measurement, as well as to more subjective evaluation. You should concentrate on inspection results and accident data and statistics when making your analysis. When you have placed into motion every chapter of this book, your inspection reports should show a decreasing number of hazards as your program continues. When this happens, your injuries and illnesses will be practically eliminated and you will be on your way to achieving greater profits for your company.

Appendix 1 USEFUL CONTACTS

SOURCES OF HELP

- American Conference of Governmental Industrial Hygienists (Ventilation and TLVs)
  600 Glenwood Avenue, Bldg. D-5, Cincinnati, OH 45211
- American Industrial Hygiene Association
  475 Wolf Ledges Parkway, Akron, OH 44311
- American National Red Cross
  17th and D Street N.W., Washington, DC 20006
- American National Standards Institute (Safety Standards)
  1430 Broadway, New York, NY 10018
- American Society of Safety Engineers
  1800 E. Oakton Street, Des Plaines, IL 60018
- American Society for Testing and Materials
  1916 Race Street, Philadelphia, PA 19103
- A.M. Best Company (Safety Directory)
  Oldwick, NJ 08858
- Board of Certified Safety Professionals
  Suite 101, 501 S. 6th Street, Champaign, IL 61820
- Bureau of Labor (Statistics)
  Room 1539 GAO, 441 G Street N.W., Washington, DC 20210
- Chemical Manufacturers Association
  2501 M Street N.W., Washington, DC 20037
- Chemtree Hot Line (Chemical Spill)
  Tel. 800-424-9300
- Factory Mutual System
  1151 Boston-Providence Turnpike, Norwood, MA 02062
- Illuminating Engineering Society (Lighting)
  345 East 47th Street, New York, NY 10017
- Industrial Hygiene News (Magazine)
  8650 Babcock Boulevard, Pittsburg, PA 15237
- Industrial Medical Association
  150 North Wacker Drive, Chicago, IL 60606
- Industrial Safety Equipment Association
  1901 N. Moore Street, Arlington, VA 22209

- International Hazard Control Managers (Certification Board)
  P.O. Box 50101, Washington, DC 20004
- National Fire Protective Association
  Batterymarch Park, Quincy, MA 02269
- National Institute for Occupational Safety and Health
  4676 Columbia Parkway, Cincinnati, OH 45226
- National Safety Council (Magazines, Books, etc.)
  444 N. Michigan Avenue, Chicago, IL 60611
- National Safety Management Society
  6060 Duke Street, Alexandria, VA 22302
- National Society for the Prevention of Blindness
  79 Madison Avenue, New York, NY 10016
- Occupational Hazards (Magazine)
  1111 Chester Avenue, Cleveland, OH 44114
- Occupational Health and Safety (Magazine)
  P.O. Box 7573, Waco, TX 76714
- Occupational Safety and Health Review Commission
  1925 K Street, N.W., Washington, DC 20006
- Superintendent of Documents
  U.S. Government Printing Office, Washington, DC 20402
- System Safety Society
  P.O. Box A, Newport Beach, CA 92663
- Underwriters Laboratories, Inc.
  207 E. Ohio Street, Chicago, IL 60611
- U.S. Department of Labor (OSHA)
  14th Street and Constitution Avenue, N.W., Washington, DC 20210
- Veterans of Safety
  4721 Briarbend Drive, Houston, TX 77035

Appendix 2   USEFUL INFORMATION

PREVENTION OF TYPICAL ACCIDENTS

Accident Categories

The following types of accidents are typical of those that occur in industry and in the Federal establishment:

1. Struck by:
   (a) Normally moving objects
   (b) Normally stationary objects
2. Contacted by
3. Struck against
4. Contact with
5. Caught between
6. Caught on
7. Caught in
8. Falls
   (a) Same level
   (b) To below
9. Over-exertion — strain accident
10. Exposure to gases, fumes, and vapors
11. Exposure accidents from other causes
12. Foreign body eye accidents

Typical Engineering — Administrative Controls

Following, in summary, are some of the engineering steps that can be taken to prevent the above types of accidents:

1. Struck by Normally Moving Objects
   (a) Install crossovers or crossunders.
   (b) Put up a barrier.
   (c) Use path markings.
   (d) Improve general illumination.

    (e) Install warning devices.

    (f) Install machine guards.

    (g) Eliminate blind corners.

    (h) Use frequent reminders.

    (i) Use a safety observer or lockout where practical.

2. Struck by Moving Objects That Leave Their Normal Paths

    (a) Permit only qualified operators.

    (b) Require periodic physical examinations for mobile equipment operators.

    (c) Check frequently for fitness to operate.

    (d) Check frequently against reckless operation.

    (e) Require regular equipment inspection.

    (f) Use protective barriers against traffic.

    (g) Study traffic patterns and establish safest routes.

    (h) Use machinery or equipment guards.

    (i) Be alert to all moving equipment in the work area.

3. Contacted by (Manual Handling)

    (a) Use safer substitute materials.

    (b) Eliminate direct manual handling.

    (c) Emphasize the use of approved containers only.

    (d) Use the best container possible.

    (e) Emphasize inspection of containers.

    (f) Require use of personal protective equipment.

    (g) Establish and enforce safe container-handling methods.

    (h) Use only approved methods, chemicals, and equipment, for cleaning.

4. Contacted by (Pressure Equipment)

    (a) Establish and enforce safe operating procedures.

    (b) Repair minor leaks promptly.

    (c) Inspect pressurized and hydraulic equipment regularly.

5. Struck against

    (a) Use boundary markings.

    (b) Inspect walkways and work areas regularly.

    (c) Provide adequate illumination.

    (d) Use warning devices.

    (e) Rope off stored materials.

    (f) Emphasize "do not run" rules.

    (g) Eliminate protruding objects.

    (h) Check new installations.

    (i) Remove obstacles.

    (j) Use attention-attracting devices.

    (k) Promote good housekeeping practices.

    (l) Improve storage arrangements.

    (m) Clean house periodically.

    (n) Prevent "stuck" conditions.

    (o) Emphasize how to handle "stuck" conditions.

    (p) Improve equipment layout.

    (q) Eliminate jobs requiring great "push-pull" operations.

    (r) Emphasize inspection of tools and equipment before use.

    (s) Emphasize use of right tool for the job.

    (t) Emphasize the principle of "controlled force."

6. Contact with

    (a) Use only authorized personnel for electrical repairs.

    (b) Use experienced men only.

    (c) Be sure electrical equipment is de-energized.

    (d) Emphasize the need to keep equipment de-energized.

    (e) Require the use of personal protective equipment.

    (f) Emphasize insulating energized parts.

    (g) Emphasize using ground insulating mats.

    (h) Emphasize using barriers.

    (i) Develop a proper respect for low voltage current.

    (j) Emphasize grounding electrical hand-tools.

    (k) Emphasize that all conductors should be regarded as hot until de-energized and checked.

    (l) Install guards.

    (m) Post warning signs.

    (n) Emphasize immediate cleanup of chemical spills.

    (o) Emphasize hand protection.

    (p) Emphasize prompt action after contact.

    (q) Dispose of acids and other chemicals properly.

7. Caught between

    (a) Install guards around moving parts.

    (b) Require equipment to be shut down before servicing.

    (c) Require that shut down equipment be locked out.

    (d) Emphasize apparel hazards around machinery.

8. Caught on

    (a) Teach workers not to create hazardous projections.

    (b) Conduct regular housekeeping inspections.

    (c) Eliminate "permanent" projections.

    (d) Use guards for "permanent" projections when they cannot be eliminated.

    (e) Use contrasting, high visibility paints.

    (f) Campaign against unsafe clothing.

    (g) Insist on correct carrying methods.

    (h) Shut down machines for repair and service work.

    (i) Require workers to lock out machines shut down for repair.

    (j) Establish safe start-up procedures.

    (k) Guard moving machinery parts.

    (l) Eliminate hazardous projections on mobile equipment.

    (m) Inspect equipment and machinery guards regularly.

    (n) Emphasize safe clothing for working around machinery or equipment.

    (o) Emphasize that distance means safety.

9. Caught in

    (a) Use inside-opening doors and hatches.

    (b) Emphasize the precaution of an "outside" safety observer.

    (c) Prohibit blocking or tying down exit covers.

    (d) Post warnings on self-locking doors.

    (e) Inspect for floor openings periodically.

    (f) Require use of barricades around temporary openings.

    (g) Improve general illumination.

    (h) Select small workers for tight places.

    (i) Emphasize pre-entry precautions.

    (j) Improve small opening access.

    (k) Warn men to call for help when caught.

10. Falls, Same Level

    (a) Inspect walkways and floor areas regularly.

    (b) Use nonskid flooring.

    (c) Emphasize prompt cleanup of spills.

    (d) Correct chronic trouble spots.

    (e) Promote the use of safe work shoes.

    (f) Provide good lighting.

    (g) Instruct men to look where they walk.

    (h) Provide adequate storage facilities for materials to prevent tripping.

    (i) Provide adequate cleanup and disposal equipment.

    (j) Emphasize cleanup during as well as after repair, construction, or service work.

    (k) Run cables overhead instead of underfoot.

    (l) Enforce the basic rules of good housekeeping.

11. Falls to Below

    (a) Emphasize check-out of ladders, etc., before use.

    (b) Emphasize safe placement of supporting equipment (staging).

    (c) Inspect supporting equipment periodically.

    (d) Teach men how to work safely from supporting equipment.

    (e) Require toe boards and back rails.

    (f) Require use of safety belts.

    (g) Use safety nets.

    (h) Use barriers for floor openings.

    (i) Use warning devices.

    (j) Use a "safety man" near openings.

    (k) Use safety belts near openings.

    (l) Use false bottoms, nets, etc., at openings.

    (m) Emphasize tripping and slipping hazards near openings.

12. Overexertion-Strain

    (a) Do not use men who are physically limited.

    (b) Emphasize safe handling methods.

    (c) Eliminate heavy manual handling.

    (d) Use more than one man.

    (e) Urge men to request help when needed.

    (f) Substitute mechanical force for manual force.

    (g) Emphasize substituting brains for brawn.

    (h) Prevent "stuck" situations.

    (i) Teach men how to maintain balance and control.

13. Exposure to Gases, Fumes, and Vapors
    (a) Increase awareness of the hazards in unventilated areas.
    (b) Post warning signs at entry points.
    (c) Insist on atmospheric tests prior to entry.
    (d) Require use of personal protective equipment.
    (e) Use two-man teams in unventilated enclosures.
    (f) Use safer substitutes for vapor-emitting materials.
    (g) Provide adequate ventilation before entry and during work period.
    (h) Limit evaporation of toxic vapor-emitting materials.
    (i) Inspect "gas hazard" equipment regularly.
    (j) Conduct routing atmospheric tests.
    (k) Emphasize safe equipment operation.
    (l) Inspect "low spots" regularly.
    (m) Emphasize opening precautions.
14. Exposure to Other Health Hazards
    (a) Radiation Exposure
        — Protect the skin against the UV radiation from welding.
        — Protect the eyes against the UV radiation from burning and welding.
        — Qualified personnel should prescribe precautions for exposure to radioactive materials.
    (b) High and Low Temperature Exposure
        — Protect employees by cooling with air moving equipment, radiant heat barriers, and limiting periods of exposure.
        — Provide adequate clothing to protect against cold.
    (c) Noise Exposure
        — Isolate noise source.
        — Enclose noise source.
        — Install mufflers on noisy equipment.
        — Use noise absorbing materials.
        — Make engineering revisions to the noise source.
        — Use administration means such as rotation of job assignments to reduce exposure time.
        — Provide personal ear protection if necessary after engineering and administrative steps have been taken.
    (d) Foreign Body, Eye
        — Provide adequate and appropriate eye protection in hazardous operations.
        — Exercise good housekeeping practices.
        — Perform prompt clean up of dusty work operations.
        — Sweep roadways adjacent to walkways frequently.
        — Cover small sand or coal stock piles if passing employees are exposed.

Preventive Maintenance

In order for a preventive maintenance program to be effective both for production and for safety and health, it must include:

- Regularly scheduled inspections of equipment, machines and tools, and of test and maintenance devices for these machines and tools;
- Regularly scheduled overhaul program for renewal of parts showing determined wear limits;
- Regularly scheduled lubrication, cleaning, etc., of machinery and equipment;
- Regularly scheduled testing of machinery and equipment for determination of performance and life cycle;
- Program for rotating machinery, equipment, and powered tools to prolong their life cycle and to provide down time for inspection, maintenance, and repair; and
- Program for equipment and tool replacement consistent with operational needs. Improvement of each factor in the total preventive maintenance program should be kept in mind when new purchases of equipment, machines, tools, and test devices are made.

SAFETY RULE BREAKERS

Dealing with safety rule breakers requires knowledge of the different types of human behavior and how one should proceed to change poor attitudes regarding safety by knowing each type and how to approach them.

Type	Approach
Sensitive	Be soft-spoken. Take time to listen and explain.
Slow	Be patient. Talk slowly and repeat several times.
Timid	Avoid being overbearing. Be calm. Try to encourage.
Careless	Arouse interest. Be firm and try to educate.
Bold	Avoid arguments. Keep temper in check. Do not flatter.
Lazy	Spark interest. Appeal to self respect.
Stubborn	Do not argue. Emphasize goals and be consistent.
Smart Aleck	Be firm. Consult person's immediate supervisor.

HAZARDOUS LOCATIONS (ELECTRICAL)

The purpose of this table is to cover the requirements for electrical equipment and wiring in locations which are classified, depending on the properties of the flammable vapors, liquids or gases, or combustible dusts or fibers which may be present. Each of the following classes is divided into two hazard categories, Division 1 and Division 2. More information regarding this subject is found in OSHA, Part 1910.399, Subpart S, Title "Electrical".

Summary of Class I, II, III Hazardous Locations

Classes	Groups	Divisions	
		1	2
I Gases, vapors and liquids (ART. 501)	A: Acetylene B: Hydrogen etc. C: Ether, etc. D: Hydrocarbons, fuels, solvents, etc.	Normally explosive and hazardous	Not normally present in an explosive concentration (but may accidentally exist)
II Dusts (ART. 502)	E: Metal dusts (conductive[a] and explosive) F: Carbon dusts (Some are conductive,[a] and all are explosive) G: Flour, starch, grain, combustible plastic or chemical dust (explosive)	Ignitable quantities of dust normally is or may be in suspension or conductive dust may be present	Dust not normally suspended in an ignitible concentration (but may accidentally exist). Dust layers are present
III Fibers and flyings (ART. 503)	Textiles, woodworking etc. (easily ignitable, but not likely to be explosive)	Handled or used in manufacturing	Stored or handled in storage (exclusive of manufacturing)

[a] Electrically conductive dusts are dusts with a resistivity less than 10^5 Ω cm.

Table courtesy of OSHA.

ELECTRICAL SHOCK

Type of Resistance	Resistance Values
Dry skin	100,000 to 600,000 Ω
Wet skin	1,000 Ω
Hand-to-foot	400 to 600 Ω
Ear-to-ear	About 100 Ω

Effects of 60-Hz Current on an Average Person

Current Through Body	Effects
1 mA or less	Causes no sensation.
1 to 8 mA	Sensation of shock. Not painful.
8 to 15 mA	Painful shock, person can let go (unsafe current).
15 to 20 mA	Painful shock, person cannot let go (unsafe current).
20 to 50 mA	Painful. Severe muscular contractions. Difficulty breathing.
100 to 200 mA	Ventricular fibrillation (a heart condition that results in death).
200 mA and over	Severe burns, severe muscular contractions, heart stops. Prevents ventricular fibrillation.

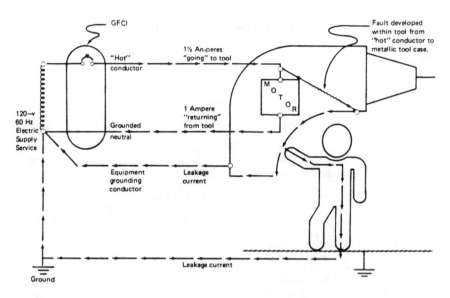

GFCI monitors the difference in current flowing into the "hot" and out to the grounded neutral conductors. The difference ($1/2$ ampere in this case) will flow back through any available path, such as the equipment grounding conductor, and through a person holding the tool, if the person is in contact with a grounded object. Figure courtesy of OSHA.

LIST OF FREQUENTLY USED ANSI STANDARDS

TITLE	ANSI NO.
Safety Requirements for Scaffolding	A10.8 — 1977
Safety Nets Used During Construction and Industrial	A10.11 — 1979
Safety Belts, Harnesses, Lanyards, Lifelines, Droplines for Construction and Industrial	A10.14 — 1975
Safety Requirements for Temporary Floor and Wall Openings, Flat Roofs, Stairs, Railings, and Toeboards for Construction	A10.18 — 1983
Safety Requirements for the Use, Care, and Protection of Abrasive Wheels	B7.1 — 1978
Safety Requirements for the Construction, Care and Use of Mechanical Power Presses	B11.1 — 1982
Safety Requirements for the Construction, Care and Use of Power Press Brakes	B11.3 — 1982
Safety Requirements for the Construction, Care and Use of Shears	B11.4 — 1983
Safety Requirements for the Construction, Care and Use of Lathes	B11.6 — 1984
Safety Requirements for the Construction, Care and Use of Drilling, Milling, and Boring Machines	B11.8 — 1983
Safety Requirements for the Construction, Care and Use of Grinding Machines	B11.9 — 1975
Safety Requirements for the Construction, Care and Use of Metal Sawing Machines	B11.10 — 1983
Floor and Wall Openings	A12.1 — 1973
Safety Requirements for Portable Wood Ladders	A14.1 — 1982
Safety Requirements for Portable Metal Ladders	A14.2 — 1982
Safety Requirements for Fixed Ladders	A14.3 — 1984

LIST OF FREQUENTLY USED ANSI STANDARDS (continued)

TITLE	ANSI NO.
Specifications for Making Buildings and Facilities Accessible to and Usable by Physically Handicapped People	A117.1 — 1980
Safety Code for Portable Air Tools	B186.1 — 1975
Fundamentals Governing Design and Operation of Local Exhaust Systems	Z9.2 — 1979
Ventilation and Safe Practices of Abrasive Blasting Operations	Z9.4 — 1979
Accident Prevention Signs, Specifications for	Z35.2 — 1968
Accident Prevention Tags, Specifications for	Z35.1 — 1972
Safety Standards for Safety Toe Footwear	Z41.1 — 1983
Safety Requirements for the Lock Out/Tag Out of Energy Sources	Z244.1 — 1982
Safety in Welding and Cutting	Z49.1 — 1983
Safety Color Code for Marking Physical Hazards	Z53.1 — 1979
Occupational Eye and Face Protection	Z87.1 — 1979
Respiratory Protection	Z88.2 — 1980
Safety Requirements for Industrial Head Protection	Z89.1 — 1981
Head Protection	Z90.1 — 1984
Safety Requirements for Working in Tanks and Other Confined Spaces	Z117.1 — 1977

COLOR CODING

RED	Means stop at red lights. Used on stop buttons. Identifies fire protection equipment. "DANGER"
ORANGE	Used on dangerous parts of equipment such as pulleys, chains, inside of guards on mechanical and electrical equipment.
YELLOW	To mark aisles, handrails, quardrails, barricades, low beams, pipes, projections, etc. "CAUTION".
GREEN	To identify first aid and safety equipment, stretchers, and safety bulletin boards.
BLUE	Used as basic color for designating machine and equipment controls such as electrical controls, valves, brakes, and disconnects.
PURPLE	Used for designating radiation hazards on warning signs, containers, and signal lights.
BLACK OR WHITE	Or a combination of black and white shall be the basic color of housekeeping markings such as, stairways (risers), location of refuse cans, food dispensing equipment, and traffic.

Table adapted from OSHA, Part 1910.144, Safety Color Code For Marking Physical Hazards.

USEFUL CONVERSION FACTORS

Weight, Area, Volume

1 cu ft water = 62.4 lb @ 60 F
1 US gal = 8.34 lb water @ 60 F
1 US gal = 231 cu in = 0.134 cu ft = 0.833 Imp gal
1 Imp gal = 277.4 cu in = 1.2 US gal
1 lb = 453.6 grams (g) = 0.454 kilograms (kg)
1 kg = 1000 g = 2.2 lb
1 cu ft = 1728 cu in = 7.48 US gal = 6.23 Imp gal
1 slug = 32.2 lb

Work and Power

1 hp = 0.745 kw = 42.4 Btu/min = 2544 Btu/hr = 33,000 ft lb/min
1 boiler hp (bhp) = 33,475 Btu/hr
1 kw = 1000 watts (w) = 1.341 hp = 56.88 Btu/min = 3412 Btu/hr
1 kw hr = 1000 w hr = 3412 Btu
1 Btu = 0.029 kw hr = 778 ft lb = 0.555 pcu (lb C unit)

Pressure and Flow

1 normal atmosphere = 1.0332 kg per sq cm.
                              = 1.0133 bars = 14.696 lb per sq in

1 lb/sq in = 2.04 in Hg @ 62 F = 2.31 ft water @ 62 F
1 lb per sq in = 0.0703 kg per sq cm
1 psia = psig + 14.7
1 kg per sq cm = 14.223 lb per sq in
1 kg per sq m = 0.2048 lb per sq ft
1 lb per sq ft = 4.8824 kg per sq m
1 kg per sq cm = 0.9678 normal atmosphere
g/cm^3 = sp gr
sp gr × 62.4 = lb/cu ft
1 gpm = 0.134 cu ft/min = 500 lb/hr × sp gr
1 cu ft/min (cfm) = 448.8 gal/hr (gph)
1 centipoise = 2.42 lb/ft hr
1 lb/ft sec = 1488 centipoises = 3600 lb/ft hr

Heat Transfer

1 Btu/hr ft^2 F = 0.0001355 g-cal sec cm^2 C
1 g-cal/sec cm^2 C = 7380 Btu/hr ft^2 F

Thermal Conductivity

1 Btu/hr ft^2 F = 0.00413 g-cal/sec cm^2 C cm
1 g-cal/sec cm^2 C cm = 242 Btu/hr ft^2 F ft

TABLES OF WEIGHTS, MEASURES AND VALUES

LONG MEASURE
United States Standard

12 inches	1 foot
3 feet	1 yard
5½ yards, or 16½ feet	1 rod
320 rods, or 5,280 feet	1 mile
1,760 yards	1 mile
40 rods	1 furlong
8 furlongs	1 statute mile
3 miles	1 league

SQUARE MEASURE
United States Standard

144 square inches	1 square foot
9 square feet	1 square yard
30¼ square yards	1 square rod
272¼ square feet	1 square rod
40 square rods	1 rood
4 roods	1 acre
160 square rods	1 acre
640 acres	1 square mile
43,560 square feet	1 acre
4,840 square yards	1 acre

SOLID OR CUBIC MEASURE (VOLUME)
United States Standard

1,728 cubic inches	1 cubic foot
27 cubic feet	1 cubic yard
128 cubic feet	1 cord of wood
24¾ cubic feet	1 perch of stone
2,150.42 cubic inches	1 standard bushel
231 cubic inches	1 standard gallon
40 cubic feet	1 ton (shipping)

DRY MEASURE
United States Standard

2 pints	1 quart
8 quarts	1 peck
4 pecks	1 bushel
2,150.42 cubic inches	1 bushel
1.2445 cubic feet	1 bushel

GEOMETRIC FORMULAS

To Obtain:	Multiply:	by:
Annulus		
Area of	Diff. of sq. of diam	0.78540
"	Diff. of sq. of radii	3.1416
Circle		
Area of	Circum.	$\frac{1}{2}$ × radius
"	Circum.	$\frac{1}{4}$ × diameter
"	Circum.^{2}.	0.079577
"	Diameter^2	0.78540
"	Radius^2.	3.1416
Circum. of	Diameter	3.1416
"	Radius	6.2832
Diam. of	Circum.	0.31831
Radius of	Circum.	0.15915
Side of equal square of . . .	Circum. of circle	0.28207
" . . .	Diam. of circle	0.88614
Side of inscribed equal. triangle of	Diam. of circle	0.86603
Side of inscribed square of	Circum. of circle	0.22508
"	Diam. of circle	0.70711
Cone, Regular		
Volume of	Area of base	$\frac{1}{3}$ × altitude
Cone, Right Circular		
Lateral area of	Radius of base	3.1416 × slant height
Volume of	(Radius of base)^2	1.0472 × altitude
Cube		
Diagonal of.	Length of one side	1.7321
Total surface area of	Area of one side	6
Volume of	Area of one side	Length of one side
"	(Length of one side)^3	1
Cycloid		
Area of	(Radius of circle)^2	9.4248
Length of arc of	Radius of circle	8
Cylinder, Hollow		
External surface area of .	External radius	6.2832 × height
Internal surface area of . .	Internal radius	6.2832 × height
Volume of	Diff. of sq. of radii	3.1416 × height
Cylinder, Truncated Right Circular		
Lateral area of	Perimeter of base	Average height
Volume of	Area of base	Average height
Ellipse		
Area of	Product of axes	0.78540
"	Product of semi-axes . . .	3.1416
Circum. of	Sum of 2 diameters	$\frac{1}{2}$ × 3.1416

MARINE WEIGHTS AND MEASURES

5280	feet	=	1 mile
16½	feet	=	1 rod
2	yards	=	1 fathom
1.15156	miles	=	1 knot
1000	millimeters	=	1 meter
100	centimeters	=	1 meter
1000	meters	=	1 kilometer
0.3937	inch	=	1 centimeter
39.37	inches	=	1 meter
25.4	millimeters	=	1 inch
.6214	mile	=	1 kilometer
144	sq. in.	=	1 sq. ft.
4840	sq. yds.	=	1 acre
231	cu. in.	=	1 gallon
32	ounces (volume)	=	1 quart
42	gallons	=	1 bbl.
1.2	U.S. gallons	=	1 Imperial gallon
1000	cubic centimeters	=	1 liter
3785	cubic centimeters	=	1 gallon
61.023	cubic inches	=	1 liter
1.0567	quarts	=	1 liter
16	ounces (weight)	=	1 pound
2000	pounds	=	1 ton — net
2240	pounds	=	1 ton — gross
2204.6	pounds	=	1 metric ton
1000	milligrams	=	1 gram
1000	grams	=	1 kilogram
453.6	grams	=	1 pound
8.328	pounds water	=	1 U.S. gallon
10	pounds water	=	1 Imperial gallon
62.4	pounds water	=	1 cu. ft.
.433	lbs. per sq. in.	=	1 foot water
.491	lbs. per sq in.	=	1 inch mercury
13.61	inches water	=	1 inch mercury
14.7	lbs. per sq .in.	=	1 atmosphere
0.0335	kgs. per sq. in.	=	1 atmosphere
.0703	kgs. per sq. in.	=	1 lb. per sq. in.
.0807	lbs. air at 32°F.	=	1 cu. ft.
550	ft. lbs. per sec.	=	1 Horsepower
745.7	Watts	=	1 Horsepower
1.34	Horsepower	=	1 Kilowatt
778	ft. lbs.	=	1 B.T.U.
2546.5	B.T.U.	=	1 H.P. Hour
1.8	B.T.U. per lb.	=	1 calorie per kg.

NOMOGRAPHIC CHART
SPECIFIC GRAVITY — POUNDS — GALLONS

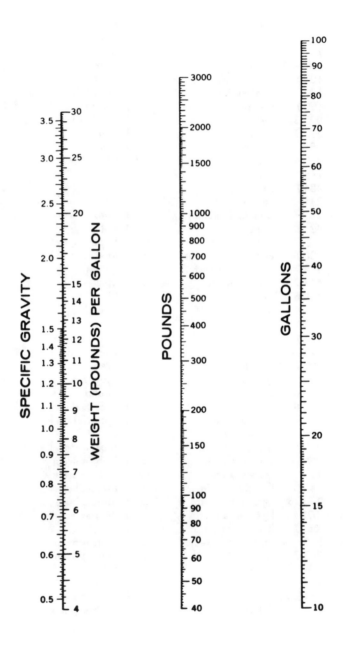

TEMPERATURE CONVERSION TABLES

INTERPOLATION FACTORS

C.		F.	C.		F.
0.56	1	1.8	3.33	6	10.8
1.11	2	3.6	3.89	7	12.6
1.67	3	5.4	4.44	8	14.4
2.22	4	7.2	5.00	9	16.2
2.78	5	9.0	5.56	10	18.0

—100 to 95

C.		F.	C.		F.
−73.3	−100	−148	6.11	43	109.4
−67.8	−90	−130	6.67	44	111.2
−62.2	−80	−112	7.22	45	113.0
−56.7	−70	−94	7.78	46	114.8
−51.1	−60	−76	8.33	47	116.6
−45.6	−50	−58	8.89	48	118.4
−40.0	−40	−40	9.44	49	120.2
−34.4	−30	−22	10.0	50	122.0
−28.9	−20	−4	10.6	51	123.8
−23.3	−10	14	11.1	52	125.6
−17.8	0	32	11.7	53	127.4
−17.2	1	33.8	12.2	54	129.2
−16.7	2	35.6	12.8	55	131.0
−16.1	3	37.4	13.3	56	132.8
−15.6	4	39.2	13.9	57	134.6
−15.0	5	41.0	14.4	58	136.4
−14.4	6	42.8	15.0	59	138.2
−13.9	7	44.6	15.6	60	140.0
−13.3	8	46.4	16.1	61	141.8
−12.8	9	48.2	16.7	62	143.6
−12.2	10	50.0	17.2	63	145.4
−11.7	11	51.8	17.8	64	147.2
−11.1	12	53.6	18.3	65	149.0
−10.6	13	55.4	18.9	66	150.8
−10.0	14	57.2	19.4	67	152.6
−9.44	15	59.0	20.0	68	154.4
−8.89	16	60.8	20.6	69	156.2
−8.33	17	62.6	21.1	70	158.0
−7.78	18	64.4	21.7	71	159.8
−7.22	19	66.2	22.2	72	161.6
−6.67	20	68.0	22.8	73	163.4
−6.11	21	69.8	23.3	74	165.2
−5.56	22	71.6	23.9	75	167.0
−5.00	23	73.4	24.4	76	168.8
−4.44	24	75.2	25.0	77	170.6
−3.89	25	77.0	25.6	78	172.4
−3.33	26	78.8	26.1	79	174.2
−2.78	27	80.6	26.7	80	176.0
−2.22	28	82.4	27.2	81	177.8
−1.67	29	84.2	27.8	82	179.6
−1.11	30	86.0	28.3	83	181.4
−0.56	31	87.8	28.9	84	183.2
−0	32	89.6	29.4	85	185.0
0.56	33	91.4	30.0	86	186.8
1.11	34	93.2	30.6	87	188.6
1.67	35	95.0	31.1	88	190.4
2.22	36	96.8	31.7	89	192.2
2.78	37	98.6	32.2	90	194.0
3.33	38	100.4	32.8	91	195.8
3.89	39	102.2	33.3	92	197.6
4.44	40	104.0	33.9	93	199.4
5.00	41	105.8	34.4	94	201.2
5.56	42	107.6	35.0	95	203.0

96 to 1100

C.		F.	C.		F.
35.6	96	204.8	304	580	1076
36.1	97	206.6	310	590	1094
36.7	98	208.4	316	600	1112
37.2	99	210.2	321	610	1130
37.8	100	212.0	327	620	1148
38	100	212	332	630	1166
43	110	230	338	640	1184
49	120	248	343	650	1202
54	130	266	349	660	1220
60	140	284	354	670	1238
66	150	302	360	680	1256
71	160	320	366	690	1274
77	170	338	371	700	1292
82	180	356	377	710	1310
88	190	374	382	720	1328
93	200	392	388	730	1346
99	210	410	393	740	1364
100	212	413	399	750	1382
104	220	428	404	760	1400
110	230	446	410	770	1418
116	240	464	416	780	1436
121	250	482	421	790	1454
127	260	500	427	800	1472
132	270	518	432	810	1490
138	280	536	438	820	1508
143	290	554	443	830	1526
149	300	572	449	840	1544
154	310	590	454	850	1562
160	320	608	460	860	1580
166	330	626	466	870	1598
171	340	644	471	880	1616
177	350	662	477	890	1634
182	360	680	482	900	1652
188	370	698	488	910	1670
193	380	716	493	920	1688
199	390	734	499	930	1706
204	400	752	504	940	1724
210	410	770	510	950	1742
216	420	788	516	960	1760
221	430	806	521	970	1778
227	440	824	527	980	1796
232	450	842	532	990	1814
238	460	860	538	1000	1832
243	470	878	543	1010	1850
249	480	896	549	1020	1868
254	490	914	554	1030	1886
260	500	932	560	1040	1904
266	510	950	566	1050	1922
271	520	968	571	1060	1940
277	530	986	577	1070	1958
282	540	1004	582	1080	1976
288	550	1022	588	1090	1994
293	560	1040	593	1100	2012
299	570	1058			

NOTE:—The numbers in bold face type refer to the temperature either in degrees Centigrade or Fahrenheit which it is desired to convert into the other scale. If converting from Fahrenheit degrees to Centigrade degrees the equivalent temperature will be found in the left column, while if converting from degrees Centigrade to degrees Fahrenheit, the answer will be found in the column on the right.

POWER PRESS GUARDING REQUIREMENTS

DISTANCE OF OPENING FROM POINT OF OPERATION HAZARD (INCHES)	MAXIMUM WIDTH OF OPENING (INCHES)
1/2 to 1 1/2	1/4
1 1/2 to 2 1/2	3/8
2 1/2 to 3 1/2	1/2
3 1/2 to 5 1/2	5/8
5 1/2 to 6 1/2	3/4
6 1/2 to 7 1/2	7/8
7 1/2 to 12 1/2	1 1/4
12 1/2 to 15 1/2	1 1/2
15 1/2 to 17 1/2	1 7/8
17 1/2 to 31 1/2	2 1/8

Note: This table shows the distance that guards shall be positioned from the danger line in accordance with the required openings.

Table courtesy of OSHA.

SAFETY LEGAL TERMS

Assumption of risk	Knowledge of an obviously dangerous condition or situation.
Burden of proof	The duty of a person to substantiate an allegation or issue.
Care	The degree of care exercised by a prudent person towards others.
Cause	That which effects a result such as an accident and/or injury.
Citation	An OSHA writ similar to a summons.
Claim	The assertion of a right regarding an injury or damage.
Contribution	The sharing by another person of the responsibility for injury to a third person.
Disability	A state of being not fully capable of performing all physical or mental functions.
Duty	The obligation of one person to another to avoid negligent injury.
Good faith	An honest intention to fulfill one's obligation.
Hazard	A risk associated with a particular type of job or workplace.
Liability	An obligation to rectify or recompense any injury or damage.
Negligence	Failure to exercise a degree of care.
Posting	To place a notice or order on display for others to see.
Reckless	Inattention to duty or indifference to consequences.
Tort	A wrongful act or failure to exercise due care.

Figure courtesy of OSHA

MATHEMATICAL INFORMATION

- To find pressure in pounds per square inch of a column of water, multiply the height of the column in feet by .434.
- Steam rising from water at its boiling point (212°F) has pressure equal to the atmosphere (14.7 lb. to the square inch).
- To find the capacity of a tank in gallons: square the diameter (inches), multiply by the length (inches), and multiply by .0034.
- A gallon of water (U.S.) weighs 8 1/3 lb., and contains 231 in^3.
- To find circumference of a circle, multiply diameter by 3.1416.
- A cubic foot of water contains 7 1/2 gallons and weighs 62 1/2 lb.
- To find diameter of a circle, multiply circumference by .31831.
- To find area of circle, multiply square of diameter by .7854.
- Doubling the diameter of a circle, increases its area four times.
- To find area of a rectangle, multiply length by width.
- To find area of a triangle, multiply base by 1/2 perpendicular height.
- The area of an ellipse equals product of both diameters × .7854.
- The area of a parallelogram equals base × altitude.
- Doubling the diameter of a pipe increases its capacity four times.

RIGHT-ANGLED TRIANGLE
To find the perpendicular height when the base and the sum of the perpendicular and hypotenuse are known

RULE:—From the square of the sum of perpendicular and hypotenuse subtract the square of the base and divide the difference by twice the sum of perpendicular and hypotenuse.

Or, divide the square of the base by the combined length of perpendicular and hypotenuse; one-half of the difference between the quotient and the combined length is the perpendicular.

EXAMPLE:—A 90-foot flagpole is broken so that its top reaches the ground 30 feet from the stump, forming a right-angled triangle. How high up is the flagpole broken?

SOLUTION:—(90 ft. $\times$ 90 ft.) — (30 ft. $\times$ 30 ft.) = 7200 ft.^{2}
7200 ft.^{2} $\div$ 2 $\times$ 90 ft. = 40 ft.
Or: (30 ft. $\times$ 30 ft.) $\div$ 90 ft. = 10 ft.
(90 ft. — 10 ft.) $\div$ 2 = 40 ft.

HOW TO FIND SQUARE ROOT

Point off the given number into periods of two places each, starting at the units place. Find the greatest number whose square is less than the first lefthand period and place it as the first figure in the quotient. Subtract its square from the lefthand period and to the remainder annex two figures of the second period for the new dividend. Double the first figure of the quotient for a partial divisor; find by trial that number which when added to 10 times the trial divisor gives a sum which when multiplied by the number itself gives a product most nearly approaching the dividend without exceeding it. This number is the second figure of the quotient; to get the third figure subtract from the new dividend the product obtained above and annex to the difference the next period, giving a new dividend. For the new trial divisor, double the first two figures of the quotient and proceed as before.

Example: Find the square root of 3.14159.

		3.	14′	15′	90	1.772
		1				
20 + 7		2	14			
		1	89			
340 + 7			25	25		
			24	29		
3540 + 2				86	90	
				70	84	

Answer is therefore 1.772.
Courtesy of Penetone Corporation.

LIQUID MEASURE (CAPACITY)
United States Standard

4 gills	1 pint
2 pints	1 quart
4 quarts	1 gallon
31½ gallons	1 barrel
2 barrels	1 hogshead
1 gallon	231 cubic inches
7.4805 gallons	1 cubic foot
16 fluid ounces	1 pint
1 fluid ounce	1.805 cubic inches
1 fluid ounce	29.59 cubic centimeters

AVOIRDUPOIS MEASURE (WEIGHT)

(Used for weighing all ordinary substances except precious metals, jewels, and drugs)

United States Standard

27¹¹⁄₃₂ grains	1 dram
16 drams	1 ounce
16 ounces	1 pound
25 pounds	1 quarter
4 quarters	1 hundredweight
100 pounds	1 hundredweight
20 hundredweight	1 ton
2,000 pounds	1 short ton
2,240 pounds	1 long ton

TROY MEASURE WEIGHT)

(Used for weighing gold, silver, and jewels)

24 grains	1 pennyweight
20 pennyweights	1 ounce
12 ounces	1 pound

Comparison of Avoirdupois and Troy Measures

1 pound troy	5,760 grains
1 pound avoirdupois	7,000 grains
1 ounce troy	480 grains
1 ounce avoirdupois	437½ grains
1 karat, or carat	3.2 troy grains
24 karats	pure gold

NUMERALS

ROMAN NUMERALS

I	1	IX	9	XVII	17	LXX	70	D	500
II	2	X	10	XVIII	18	LXXX		DC	600
III	3	XI	11	XIX	19	or XXC	80	DCC	700
IV	4	XII	12	XX	20	XC	90	DCCC	800
V	5	XIII	13	XXX	30	C	100	CM	900
VI	6	XIV	14	XL	40	CC	200	M or	
VII	7	XV	15	L	50	CCC	300	cIc	1000
VIII	8	XVI	16	LX	60	CCCC	400	MM	2000

Note. A dash line over a numeral multiplies the value by 1,000. Thus, $\overline{X} = 10,000$; $\overline{L} = 50,000$; $\overline{C} = 100,000$; $\overline{D} = 500,000$; $\overline{M} = 1,000,000$; $\overline{CLIX} = 159,000$; $\overline{DLIX} = 559,000$.

GENERAL RULES IN ROMAN NUMERALS

(1) Repeating a letter repeats its value: XX = 20; CCC = 300.

(2) A letter placed after one of greater value adds thereto: VIII = 8; DC = 600.

(3) A letter placed before one of greater value subtracts therefrom: IX = 9; CM = 900.

ARABIC NUMERALS

Trillions	Billions	Millions	Thousands	Hundreds
7,	256,	423,	896,	384

Note: In the United States and France a billion is a thousand millions (1,000,000,000). In Britain and Germany a billion is a million millions (1,000,000,000,000).

ACIDITY AND ALKALINITY

The pH Value

The use of pH values to express the degree of acidity or alkalinity of solutions was at one time merely of theoretical interest. Now, with the availability of rugged industrial-type pH meters, it is one of the most common controls for chemical and processing industries.

A simplified explanation of the meaning of pH may be of interest to those not already familiar with it.

In the pH scale, the figure 7.0 represents an exact neutrality. This is the pH of chemically pure water. If an alkali such as caustic soda is added to pure water, the pH value of the solution is increased and may rise to values as high as 14.0. If, on the other hand, an acid such as hydrochloric acid is added, the pH value decreases as more acid is added.

The pH value, therefore, indicates first of all the condition of a solution with respect to acidity or alkalinity. For example, the fact that a solution has a pH value of 6.0 indicates that it has an acid reaction, while a pH value of 8.0 indicates an alkaline solution. These values, however, have an additional and much more important function in that they are an absolutely accurate measure of the *relative* degree of acidity or alkalinity.

For example, the value pH 6.0 means that a solution has an acidity ten times that of pure water while for pH 5.0 the value is 100 times that of water. The table below shows these and other values graphically:

pH Value	Relative Acidity or Alkalinity in Terms of Pure Water.	
0	X 10,000,000	⎫
1	X 1,000,000	⎪
2	X 100,000	⎪
3	X 10,000	⎬ Acidity
4	X 1,000	⎪
5	X 100	⎪
6	X 10	⎭
7	1	Pure Water
8	X 10	⎫
9	X 100	⎪
10	X 1,000	⎪
11	X 10,000	⎬ Alkalinity
12	X 100,000	⎪
13	X 1,000,000	⎪
14	X 10,000,000	⎭

The following example will illustrate the value of pH in determining the relative degree of acidity of two solutions:

ACIDITY AND ALKALINITY

Two acid solutions are made up, one containing 0.37% of hydrochloric acid, the other 0.60% of acetic acid. Equal quantities of these two solutions require exactly the same quantity of a solution of a caustic soda to neutralize them and, according to this method of evaluation, they are of an equal degree of acidity. Determinations of the pH values of the two solutions, however, would show:

0.37% hydrochloric acid = pH 1.04
0.60% acetic acid   = pH 2.89

Reference to the table on the preceding page will show that the difference in these two pH values indicates that the hydrochloric acid solution is nearly 100 times more strongly acid than the acetic acid solution. In this instance, it will be noted that pH values bring out clearly the fact that hydrochloric acid is a very strong acid while acetic acid is relatively weak. The general behavior of these two acids readily confirms this fact.

Equipment is now available for making rapid and accurate determinations of pH values either by colorimetric or electrometric methods. No technical training is required to use this equipment.

RELATION BETWEEN HEAT COLORS AND APPROPRIATE TEMPERATURES

Color	Temperature	
	°F	°C
Dazzling	3450	1900
White	2200	1200
Light yellow	1975	1080
Yellow	1825	995
Orange	1725	940
Salmon	1650	900
Bright red	1550	845
Cherry or full red	1375	745
Medium cherry	1175	635
Dark red	1050	565
Red just visible	930	500

THE DILUTION AND CONCENTRATION
OF LIQUIDS AND MIXTURES
RECTANGLE METHOD

The figures expressing the percentage concentration of two solutions (or those of one solution, and the figure 0 *for water,* where dilution with water is desired) are written in the two left hand corners of a rectangle, and the figure expressing the desired concentration is placed on the intersection of the two diagonals of this rectangle.

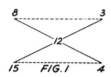

Now subtract the figures on the diagonals, and write the result at the other end of the respective diagonal. These figures then indicate what quantities of the solutions whose concentration is given on the other end of the respective *horizontal* line, must be taken to obtain a solution of the desired concentration. For example, to make a 12% solution, by mixing an 8% and a 15% solution we prepare Fig. 1, which indicates that we have to take 3 parts by weight of the 8% solution, and 4 parts by weight of the 15% solution to obtain (7 parts by weight of) the 12% solution.

Again, if we wish to dilute a 25% solution so as to obtain a 9% solution, we place the figure 25 in, for example, the upper left corner of a rectangle and place figure 0 (concentration of the solution in pure water) in the lower left corner, and then place the figure 9 (desired concentration) at the point of intersection of the diagonals, and subtracting across the diagonals, we obtain Fig. 2: 9 parts by weight of the 25% solution, if mixed with 16 parts by weight of water, will give 25 parts by weight of a 9% solution.

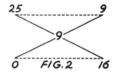

SELECTION CHART
RECOMMENDED EYE AND FACE PROTECTORS FOR USE IN INDUSTRY, SCHOOLS, AND COLLEGES

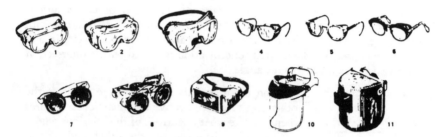

1. GOGGLES, Flexible Fitting Regular Ventilation
2. GOGGLES, Flexible Fitting, Hooded Ventilation
3. GOGGLES, Cushioned Fitting Rigid Body
*4. SPECTACLES, Metal Frame, with Sideshields
*5. SPECTACLES, Plastic Frame, with Sideshilds
*6. SPECTACLES, Metal-Plastic Frame, with Sideshields
**7. WELDING GOGGLES, Eyecup Type, Tinted Lenses (Illustrated)
7A. CHIPPING GOGGLES, Eyecup Type, Clear Safety Lenses (Not Illustrated)
**8. WELDING GOGGLES, Coverspec Type Tinted Lenses (Illustrated)
8A. CHIPPING GOGGLES, Coverspec Type, Clear Safety Lenses (Not Illustrated)
**9. WELDING GOGGLES, Coverspec Type, Tinted Plate Lens
10. FACE SHIELD (Available with Plastic or Mesh Window)
**11. WELDING HELMETS

*   Non-sideshield spectacles are available for limited hazard use requiring only frontal protection.

** See appendix chart "Selection of Shade Numbers for Welding Filters."

Applications

Operation	Hazards	Recommended Protectors[a]
Acetylene — burning Acetylene — cutting Acetylene — welding	Sparks, harmful rays molten metal, flying particles	7, **8**, **9**
Chemical handling	Splash, acid burns, fumes	**2**, 10 (For severe exposure add 10 over 2)
Chipping	Flying particles	**1**, **3**, 4, 5, 6, **7A**, **8A**
Electric (ARC) welding	Sparks, intense rays, molten metal	9, **11** (11 in combination with 4, 5, 6, in tinted lenses, advisable)
Furnace operations	Glare, heat, molten metal	7, **8**, 9 (For severe exposure add **10**)
Grinding — light	Flying particles	1, **3**, 4, **5**, **6**, 10
Grinding — heavy	Flying particles	1, **3**, **7A**, **8A** (For severe exposure add 10)
Laboratory	Chemical splash, glass breakage	**2** (10 when in combination with 4, 5, 6)
Machining	Flying particles	1, **3**, 4, **5**, **6**, 10
Molten metals	Heat, glare, sparks, splash	7, **8**, (10 in combination with 4,5,6, in tinted lenses)
Spot welding	Flying particles, sparks	1, **3**, **4**, **5**, **6**, 10

Eye protection chart from ANSI Standard Z87.1, Practice for Occupational and Educational Eye and Face Protection.

[a] Bold type numbers signify preferred protection.

SAFETY AND HEALTH FILMS

- *Little Things That Count*
- *Housekeeping Means Safekeeping*
- *Down and Out*
- *Don't Push Your Luck*
- *Bend Your Knees*
- *One Last Shock*
- *To Last a Lifetime*
- *Eye Emergency*
- *Future In Your Hands*
- *Safety Belts*
- *Excavation*
- *Scaffolding*
- *Men of Iron*
- *Right On Roofer*

Available sources

The Film Library, P.O. Box 76146, Los Angeles, CA 90076, Telephone: (213) 381-5569.

International Film Bureau, Inc., 332 S. Michigan Avenue, Chicago, IL 60604, Telephone: (312) 427-4545.

FIRST AID INJURIES

WOUNDS — An abrasion results from scraping the skin.

INCISIONS — Skin cuts caused by knives or other cutting objects.

LACERATIONS — Jagged, irregular breaks or tears of soft tissues.

PUNCTURE — Wounds produced by pointed objects such as nails.

AVULSION — Forcible separation or tearing of tissue from body.

SEVERE BLEEDING — Loss of more than a quart of blood.

BITES — Produced by animals or human bite.

CLOSED WOUND — Black eye or any massive injury to soft tissue.

EYE INJURY — Caused by dust, chemical or foreign object.

SHOCK — Caused by loss of blood, chemical poisoning, drugs, alcohol.

RESPIRATORY FAILURE — Caused by obstruction, depleted oxygen, toxic gases.

DROWNING — Aspiration of fluids.

POISONING — Injection, inhalation or absorption of a harmful substance.

BURNS — Results from heat, chemical or radiation. First Degree: Redness, mild swelling, and pain; Second Degree: Red or mottled with blisters; and Third Degree: Charring and destruction of red blood cells.

FROST BITE — Drowsiness, staggering, shock and eyesight fails.

HEAT STROKE — Muscular pains, fatigue, and weakness.

FRACTURE — (Simple bone break) no skin laceration. (Compound bone break) deep skin laceration.

SKULL FRACTURE — Caused by severe falls or flying particles.

SUDDEN ILLNESS — Caused by poisoning, asphyxia, head injury, stroke, heart attack, convulsions, epilepsy, hyperglycemia, hemorrhage, headache, glaucoma, toothache, ear infection, chest pains, abdominal pains, food poisoning, allergic reactions, acute appendicitis, hernia, gallstones and kidney stones, ulcer, back pain, bee sting, high fever, and mental and emotional disturbance.

FIRE ASSOCIATED TEMPERATURES

MATERIAL	MELTING POINT	REMARKS
Acetate	887°F	Ignites at 887°F
Aluminum	1250°F	
Benzine		Flashpoint 12°F
Brass	1600-2000°F	
Chromium	3430°F	
Engine Oil		Flashes at 400°F; ignites at 1000°F
Ethyl alcohol		Ignites at 1040°F
Fuel oil		Flashes at 100°F; ignites at 400°F
Gasoline		Flashes at 80°F; ignites at 850°F
Glass cloth		Fuses at 1200°F
Hydraulic fluid		Flashes at 200°F; ignites at 640°F
Iron	2800°F	
JP-4		Flashes at -60°F; ignites at 484°F
Magnesium	1250°F	
Neoprene		Blisters at 500°F
Nomex (fire retardant)		Ignites at 2372°F
Nylon fiber	320-500°F	
Nylon tubing	250-300°F	
Paint		Softens at 400°F; Blisters at 800°F
Silicone		Blisters at 700°F
Silver	1760°F	
Stainless steel	2700°F	Starts discoloring at 850°F
Teflon	630°F	
Tin	449°F	
Titanium	3100°F	Scale form at 110°F
Wire bundles		Outer braid brittle at 500°F
Wood		Ignites at 400-500°F
Zinc chromate		Tans at 450°F; browns at 600°F

FREE FALLING BODIES

The formula is:

$$s = \frac{g \times t^2}{2}$$

Where

s = distance in feet

g = acceleration of gravity (32 feet per second per second)

t = time in seconds

☆ U.S. GOVERNMENT PRINTING OFFICE: 1979-658-620

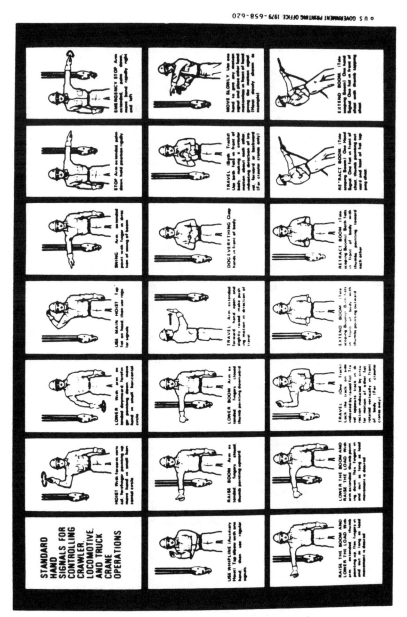

Standard hand signals for controlling crawler, locomotive and truck crane operations.

OSHA CONDITIONS OF EXPOSURE

Exposure duration (hours per 24-hour day)	Sound level (decibel-dBA)
8	90
6	92
4	95
3	97
2	100
1½	102
1	105
½	110
¼ or less	115

Example: According to OSHA regulations, a person exposed to a sound level of 102 decibels can only be exposed without hearing protectors for 1½ hours.

NOISE IN DECIBELS

Activity	Decibels
Rocket launching pad	180
Jet plane	140
Gunshot blast	140
Riveting a tank	130
Auto horn	120
Sandblasting	112
Woodworking shop	100
Punch press	100
Pneumatic drill	100
Boiler shop	100
Hydraulic press	100
Subway	90
Average factory	85
Computer room	85
Large restaurant	80
Office machines	80
Busy traffic	75
Normal speech	65
Average home	50
Quiet office	40
Soft whisper	30

Note: Noise over 85 decibels may harm hearing. Noise over 140 decibels may cause pain.

Table courtesy of OSHA.

RAMPS, STAIRS & LADDERS: PREFERRED ANGLES

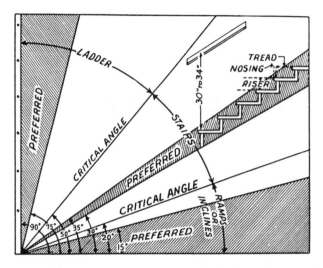

Figure courtesy of the National Safety Council.

REASONABLE WEIGHT LIMITS FOR OCCASIONAL LIFTING

AGE	MALE (lb.)	FEMALE (lb.)
14–15	33	22
16–18	42	26
18–20	51	31
20–35	55	33
35–50	46	29
over 50	35	22

From the Swiss Accident Institute.

SAFETY TRENCHING REQUIREMENTS

Banks more than 5 feet high shall be shored, laid back to a stable slope, or some other equivalent means of protection shall be provided when employees may be exposed to moving ground or cave-ins. Because cave-ins are killers, OSHA requires that, in all excavations, employees exposed to dangers from moving ground shall be protected by a shoring system, sloping of the ground, or some other equivalent means. For more information consult with OSHA, Part 1926.652, Subpart P, Titled "Excavation, Trenching and Shoring".

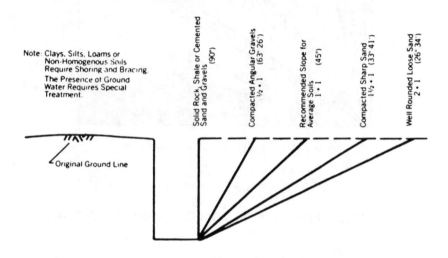

**APPROXIMATE ANGLE OF REPOSE
FOR SLOPING OF SIDES OF EXCAVATIONS**

Figure courtesy of OSHA.

WINDCHILL TABLE

MPH	wind speed	Thermometer reading (°F)										
		50	40	30	20	10	0	−10	−20	−30	−40	−50
0	no movement	wind chill same as thermometer reading if no breeze										
10	feel breeze on face	40	28	16	4	−9	−21	−33	−46	−58	−70	−83
15	moderate breeze	36	22	9	−5	−18	−36	−45	−58	−72	−85	−99
20	small branches move	32	18	4	−10	−25	−39	−53	−67	−82	−96	−110
25	strong breeze	30	16	0	−15	−29	−44	−59	−74	−88	−104	−118
30	large branches move	28	13	−2	−18	−33	−48	−63	−79	−94	−109	−125
40	gale	26	10	−6	−21	−37	−53	−69	−85	−100	−116	−132
		little danger if clothed properly			exposed flesh can be in danger here				frostbite easily here so use extreme caution			

Note: Effective temperature is found at intersection of thermometer reading and wind speed. As an example, with a wind speed of 25 and thermometer reading of 10°F an effective temperature of −29°F exists. Here exposed flesh is in danger.

If you want to go metric and convert the reading to Celsius (°C), subtract 32° from °F and multiply by 5/9. In this case you'll find that −29°F equals −34°C.

Chart sources: Dept. of Navy, Dept. of Army, Appalachian Mt. Club, U.S. Weather Bureau, National Science Foundation.

SAFETY SLOGANS

- "Accidents do not happen — but are caused."
- "Safety has no quitting time."
- "A safe plant is an efficient plant."
- "One accident is one too many."
- "The key person in safety is the foreman."
- "Be careful today; be safe tomorrow."
- "It is better to lead to safety than to drive safety."
- "The skillful worker is the safe worker."
- "Safety should be cooperative rather than enforced."
- "Take care — not chances."
- "Safety is good business."
- "The end of a perfect day — no accidents."
- "The rule breaker is an accident maker."
- "Employ safety — it works."
- "Be alert and don't get hurt."
- "Can you afford an accident?"
- "Safety's the game — prevention the aim."
- "Don't let carelessness take away your job."
- "Safety is no accidents."
- "Safety is here to stay."
- "Safety and production go hand-in-hand."
- "Let safety share your work."
- "Stop! Look! and lessen accidents."

Appendix 3

USEFUL OSHA
INFORMATION

Public Law 91-596
91st Congress, S. 2193
December 29, 1970

An Act

84 STAT. 1590

To assure safe and healthful working conditions for working men and women;
by authorizing enforcement of the standards developed under the Act; by
assisting and encouraging the States in their efforts to assure safe and healthful working conditions; by providing for research, information, education, and
training in the field of occupational safety and health; and for other purposes.

*Be it enacted by the Senate and House of Representatives of the
United States of America in Congress assembled,* That this Act may
be cited as the "Occupational Safety and Health Act of 1970".

Occupational
Safety and
Health Act of
1970.

CONGRESSIONAL FINDINGS AND PURPOSE

SEC. (2) The Congress finds that personal injuries and illnesses arising out of work situations impose a substantial burden upon, and are
a hindrance to, interstate commerce in terms of lost production, wage
loss, medical expenses, and disability compensation payments.

(b) The Congress declares it to be its purpose and policy, through
the exercise of its powers to regulate commerce among the several
States and with foreign nations and to provide for the general welfare,
to assure so far as possible every working man and woman in the
Nation safe and healthful working conditions and to preserve our
human resources—

(1) by encouraging employers and employees in their efforts
to reduce the number of occupational safety and health hazards
at their places of employment, and to stimulate employers and
employees to institute new and to perfect existing programs for
providing safe and healthful working conditions;

(2) by providing that employers and employees have separate
but dependent responsibilities and rights with respect to achieving safe and healthful working conditions;

(3) by authorizing the Secretary of Labor to set mandatory
occupational safety and health standards applicable to businesses
affecting interstate commerce, and by creating an Occupational
Safety and Health Review Commission for carrying out adjudi-
catory functions under the Act;

(4) by building upon advances already made through employer
and employee initiative for providing safe and healthful working
conditions;

(5) by providing for research in the field of occupational
safety and health, including the psychological factors involved,
and by developing innovative methods, techniques, and
approaches for dealing with occupational safety and health
problems;

(6) by exploring ways to discover latent diseases, establishing
causal connections between diseases and work in environmental
conditions, and conducting other research relating to health problems, in recognition of the fact that occupational health standards
present problems often different from those involved in occupational safety;

(7) by providing medical criteria which will assure insofar as
practicable that no employee will suffer diminished health, functional capacity, or life expectancy as a result of his work
experience;

(8) by providing for training programs to increase the number and competence of personnel engaged in the field of occupational safety and health;

84 STAT. 1591 **Pub. Law 91-596**   **- 2 -**   **December 29, 1970**

(9) by providing for the development and promulgation of occupational safety and health standards;

(10) by providing an effective enforcement program which shall include a prohibition against giving advance notice of any inspection and sanctions for any individual violating this prohibition;

(11) by encouraging the States to assume the fullest responsibility for the administration and enforcement of their occupational safety and health laws by providing grants to the States to assist in identifying their needs and responsibilities in the area of occupational safety and health, to develop plans in accordance with the provisions of this Act, to improve the administration and enforcement of State occupational safety and health laws, and to conduct experimental and demonstration projects in connection therewith;

(12) by providing for appropriate reporting procedures with respect to occupational safety and health which procedures will help achieve the objectives of this Act and accurately describe the nature of the occupational safety and health problem;

(13) by encouraging joint labor-management efforts to reduce injuries and disease arising out of employment.

DEFINITIONS

SEC. 3. For the purposes of this Act—

(1) The term "Secretary" mean the Secretary of Labor.

(2) The term "Commission" means the Occupational Safety and Health Review Commission established under this Act.

(3) The term "commerce" means trade, traffic, commerce, transportation, or communication among the several States, or between a State and any place outside thereof, or within the District of Columbia, or a possession of the United States (other than the Trust Territory of the Pacific Islands), or between points in the same State but through a point outside thereof.

(4) The term "person" means one or more individuals, partnerships, associations, corporations, business trusts, legal representatives, or any organized group of persons.

(5) The term "employer" means a person engaged in a business affecting commerce who has employees, but does not include the United States or any State or political subdivision of a State.

(6) The term "employee" means an employee of an employer who is employed in a business of his employer which affects commerce.

(7) The term "State" includes a State of the United States, the District of Columbia, Puerto Rico, the Virgin Islands, American Samoa, Guam, and the Trust Territory of the Pacific Islands.

(8) The term "occupational safety and health standard" means a standard which requires conditions, or the adoption or use of one or more practices, means, methods, operations, or processes, reasonably necessary or appropriate to provide safe or healthful employment and places of employment.

(9) The term "national consensus standard" means any occupational safety and health standard or modification thereof which (1), has been adopted and promulgated by a nationally recognized standards-producing organization under procedures whereby it can be determined by the Secretary that persons interested

and affected by the scope or provisions of the standard have reached substantial agreement on its adoption, (2) was formulated in a manner which afforded an opportunity for diverse views to be considered and (3) has been designated as such a standard by the Secretary, after consultation with other appropriate Federal agencies.

(10) The term "established Federal standard" means any operative occupational safety and health standard established by any agency of the United States and presently in effect, or contained in any Act of Congress in force on the date of enactment of this Act.

(11) The term "Committee" means the National Advisory Committee on Occupational Safety and Health established under this Act.

(12) The term "Director" means the Director of the National Institute for Occupational Safety and Health.

(13) The term "Institute" means the National Institute for Occupational Safety and Health established under this Act.

(14) The term "Workmen's Compensation Commission" means the National Commission on State Workmen's Compensation Laws established under this Act.

APPLICABILITY OF THIS ACT

Sec. 4. (a) This Act shall apply with respect to employment performed in a workplace in a State, the District of Columbia, the Commonwealth of Puerto Rico, the Virgin Islands, American Samoa, Guam, the Trust Territory of the Pacific Islands, Wake Island, Outer Continental Shelf lands defined in the Outer Continental Shelf Lands Act, Johnston Island, and the Canal Zone. The Secretary of the Interior shall, by regulation, provide for judicial enforcement of this Act by the courts established for areas in which there are no United States district courts having jurisdiction.   67 Stat. 462. 43 USC 1331 note.

(b) (1) Nothing in this Act shall apply to working conditions of employees with respect to which other Federal agencies, and State agencies acting under section 274 of the Atomic Energy Act of 1954, as amended (42 U.S.C. 2021), exercise statutory authority to prescribe or enforce standards or regulations affecting occupational safety or health.   73 Stat. 688.

(2) The safety and health standards promulgated under the Act of June 30, 1936, commonly known as the Walsh-Healey Act (41 U.S.C. 35 et seq.), the Service Contract Act of 1965 (41 U.S.C. 351 et seq.), Public Law 91-54, Act of August 9, 1969 (40 U.S.C. 333), Public Law 85-742, Act of August 23, 1958 (33 U.S.C. 941), and the National Foundation on Arts and Humanities Act (20 U.S.C. 951 et seq.) are superseded on the effective date of corresponding standards, promulgated under this Act, which are determined by the Secretary to be more effective. Standards issued under the laws listed in this paragraph and in effect on or after the effective date of this Act shall be deemed to be occupational safety and health standards issued under this Act, as well as under such other Acts.   49 Stat. 2036. 79 Stat. 1034. 83 Stat. 96. 72 Stat. 835. 79 Stat. 845; Ante, p. 443.

(3) The Secretary shall, within three years after the effective date of this Act, report to the Congress his recommendations for legislation to avoid unnecessary duplication and to achieve coordination between this Act and other Federal laws.   Report to Congress.

84 STAT. 1593 **Pub. Law 91-596**   - 4 -   **December 29, 1970**

(4) Nothing in this Act shall be construed to supersede or in any manner affect any workmen's compensation law or to enlarge or diminish or affect in any other manner the common law or statutory rights, duties, or liabilities of employers and employees under any law with respect to injuries, diseases, or death of employees arising out of, or in the course of, employment.

DUTIES

Sec. 5. (a) Each employer—
(1) shall furnish to each of his employees employment and a place of employment which are free from recognized hazards that are causing or are likely to cause.death or serious physical harm to his employees;
(2) shall comply with occupational safety and health standards promulgated under this Act.
(b) Each employee shall comply with occupational safety and health standards and all rules, regulations, and orders issued pursuant to this Act which are applicable to his own actions and conduct.

OCCUPATIONAL SAFETY AND HEALTH STANDARDS

Sec. 6. (a) Without regard to chapter 5 of title 5, United States
80 Stat. 381; Code, or to the other subsections of this section, the Secretary shall,
81 Stat. 195. as soon as practicable during the period beginning with the effective
5 USC 500. date of this Act and ending two years after such date, by rule promulgate as an occupational safety or health standard any national consensus standard, and any established Federal standard, unless he determines that the promulgation of such a standard would not result in improved safety or health for specifically designated employees. In the event of conflict among any such standards, the Secretary shall promulgate the standard which assures the greatest protection of the safety or health of the affected employees.
(b) The Secretary may by rule promulgate, modify, or revoke any occupational safety or health standard in the following manner:
(1) Whenever the Secretary, upon the basis of information submitted to him in writing by an interested person, a representative of any organization of employers or employees, a nationally recognized standards-producing organization, the Secretary of Health, Education, and Welfare, the National Institute for Occupational Safety and Health, or a State or political subdivision, or on the basis of information developed by the Secretary or otherwise available to him, determines that a rule should be promulgated in order to serve the objec-
Advisory   tives of this Act, the Secretary may request the recommendations of
committee,   an advisory committee appointed under section 7 of this Act. The Sec-
recommendations. retary shall provide such an advisory committee with any proposals of his own or of the Secretary of Health, Education, and Welfare, together with all pertinent factual information developed by the Secretary or the Secretary of Health, Education, and Welfare, or otherwise available, including the results of research, demonstrations, and experiments. An advisory committee shall submit to the Secretary its recommendations regarding the rule to be promulgated within ninety days from the date of its appointment or within such longer or shorter period as may be prescribed by the Secretary, but in no event for a period which is longer than two hundred and seventy days.

December 29, 1970   - 5 -   Pub. Law 91-596   84 STAT. 1594

(2) The Secretary shall publish a proposed rule promulgating, modifying, or revoking an occupational safety or health standard in the Federal Register and shall afford interested persons a period of thirty days after publication to submit written data or comments. Where an advisory committee is appointed and the Secretary determines that a rule should be issued, he shall publish the proposed rule within sixty days after the submission of the advisory committee's recommendations or the expiration of the period prescribed by the Secretary for such submission.   Publication in Federal Register.

(3) On or before the last day of the period provided for the submission of written data or comments under paragraph (2), any interested person may file with the Secretary written objections to the proposed rule, stating the grounds therefor and requesting a public hearing on such objections. Within thirty days after the last day for filing such objections, the Secretary shall publish in the Federal Register a notice specifying the occupational safety or health standard to which objections have been filed and a hearing requested, and specifying a time and place for such hearing.   Hearing, notice.   Publication in Federal Register.

(4) Within sixty days after the expiration of the period provided for the submission of written data or comments under paragraph (2), or within sixty days after the completion of any hearing held under paragraph (3), the Secretary shall issue a rule promulgating, modifying, or revoking an occupational safety or health standard or make a determination that a rule should not be issued. Such a rule may contain a provision delaying its effective date for such period (not in excess of ninety days) as the Secretary determines may be necessary to insure that affected employers and employees will be informed of the existence of the standard and of its terms and that employers affected are given an opportunity to familiarize themselves and their employees with the existence of the requirements of the standard.

(5) The Secretary, in promulgating standards dealing with toxic materials or harmful physical agents under this subsection, shall set the standard which most adequately assures, to the extent feasible, on the basis of the best available evidence, that no employee will suffer material impairment of health or functional capacity even if such employee has regular exposure to the hazard dealt with by such standard for the period of his working life. Development of standards under this subsection shall be based upon research, demonstrations, experiments, and such other information as may be appropriate. In addition to the attainment of the highest degree of health and safety protection for the employee, other considerations shall be the latest available scientific data in the field, the feasibility of the standards, and experience gained under this and other health and safety laws. Whenever practicable, the standard promulgated shall be expressed in terms of objective criteria and of the performance desired.   Toxic materials.

(6) (A) Any employer may apply to the Secretary for a temporary order granting a variance from a standard or any provision thereof promulgated under this section. Such temporary order shall be granted only if the employer files an application which meets the requirements of clause (B) and establishes that (i) he is unable to comply with a standard by its effective date because of unavailability of professional or technical personnel or of materials and equipment needed to come into compliance with the standard or because necessary construction or alteration of facilities cannot be completed by the effective date, (ii) he is taking all available steps to safeguard his employees against the hazards covered by the standard, and (iii) he has an effective program for coming into compliance with the standard as quickly as   Temporary variance order.

84 STAT. 1595   **Pub. Law 91-596**   **- 6 -**   **December 29, 1970**

practicable. Any temporary order issued under this paragraph shall prescribe the practices, means, methods, operations, and processes which the employer must adopt and use while the order is in effect and state in detail his program for coming into compliance with the standard. Such a temporary order may be granted only after notice to employees and an opportunity for a hearing: *Provided*, That the Secretary may issue one interim order to be effective until a decision is made on the basis of the hearing. No temporary order may be in effect for longer than the period needed by the employer to achieve compliance with the standard or one year, whichever is shorter, except that such an order may be renewed not more than twice (I) so long as the requirements of this paragraph are met and (II) if an application for renewal is filed at least 90 days prior to the expiration date of the order. No interim renewal of an order may remain in effect for longer than 180 days.

(B) An application for a temporary order under this paragraph (6) shall contain:

    (i) a specification of the standard or portion thereof from which the employer seeks a variance,

    (ii) a representation by the employer, supported by representations from qualified persons having firsthand knowledge of the facts represented, that he is unable to comply with the standard or portion thereof and a detailed statement of the reasons therefor,

    (iii) a statement of the steps he has taken and will take (with specific dates) to protect employees against the hazard covered by the standard,

    (iv) a statement of when he expects to be able to comply with the standard and what steps he has taken and what steps he will take (with dates specified) to come into compliance with the standard, and

    (v) a certification that he has informed his employees of the application by giving a copy thereof to their authorized representative, posting a statement giving a summary of the application and specifying where a copy may be examined at the place or places where notices to employees are normally posted, and by other appropriate means.

A description of how employees have been informed shall be contained in the certification. The information to employees shall also inform them of their right to petition the Secretary for a hearing.

(C) The Secretary is authorized to grant a variance from any standard or portion thereof whenever he determines, or the Secretary of Health, Education, and Welfare certifies, that such variance is necessary to permit an employer to participate in an experiment approved by him or the Secretary of Health, Education, and Welfare designed to demonstrate or validate new and improved techniques to safeguard the health or safety of workers.

(7) Any standard promulgated under this subsection shall prescribe the use of labels or other appropriate forms of warning as are necessary to insure that employees are apprised of all hazards to which they are exposed, relevant symptoms and appropriate emergency treatment, and proper conditions and precautions of safe use or exposure. Where appropriate, such standard shall also prescribe suitable protective equipment and control or technological procedures to be used in connection with such hazards and shall provide for monitoring or measuring employee exposure at such locations and intervals, and in such manner as may be necessary for the protection of employees. In

Margin notes:
Notice, hearing.
Renewal.
Time limitation.
Labels, etc.
Protective equipment, etc.

addition, where appropriate, any such standard shall prescribe the   Medical
type and frequency of medical examinations or other tests which shall examinations.
be made available, by the employer or at his cost, to employees exposed
to such hazards in order to most effectively determine whether the
health of such employees is adversely affected by such exposure. In the
event such medical examinations are in the nature of research, as determined by the Secretary of Health, Education, and Welfare, such examinations may be furnished at the expense of the Secretary of Health,
Education, and Welfare. The results of such examinations or tests
shall be furnished only to the Secretary or the Secretary of Health,
Education, and Welfare, and, at the request of the employee, to his
physician. The Secretary, in consultation with the Secretary of Health,
Education, and Welfare, may by rule promulgated pursuant to sec-   80 Stat. 383.
tion 553 of title 5, United States Code, make appropriate modifications in the foregoing requirements relating to the use of labels or
other forms of warning, monitoring or measuring, and medical examinations, as may be warranted by experience, information, or medical
or technological developments acquired subsequent to the promulgation of the relevant standard.

(8) Whenever a rule promulgated by the Secretary differs substan-   Publication
tially from an existing national consensus standard, the Secretary   in Federal
shall, at the same time, publish in the Federal Register a statement Register.
of the reasons why the rule as adopted will better effectuate the purposes of this Act than the national consensus standard.

(c)(1) The Secretary shall provide, without regard to the require-   Temporary
ments of chapter 5, title 5, United States Code, for an emergency tem- standard.
porary standard to take immediate effect upon publication in the   Publication
Federal Register if he determines (A) that employees are exposed to   in Federal
grave danger from exposure to substances or agents determined to be   Register.
toxic or physically harmful or from new hazards, and (B) that such   80 Stat. 381;
emergency standard is necessary to protect employees from such   81 Stat. 195.
danger.   5 USC 500.

(2) Such standard shall be effective until superseded by a standard   Time
promulgated in accordance with the procedures prescribed in para-   limitation.
graph (3) of this subsection.

(3) Upon publication of such standard in the Federal Register the
Secretary shall commence a proceeding in accordance with section
6(b) of this Act, and the standard as published shall also serve as a
proposed rule for the proceeding. The Secretary shall promulgate a
standard under this paragraph no later than six months after publication of the emergency standard as provided in paragraph (2) of this
subsection.

(d) Any affected employer may apply to the Secretary for a rule or   Variance rule.
order for a variance from a standard promulgated under this section.
Affected employees shall be given notice of each such application and
an opportunity to participate in a hearing. The Secretary shall issue
such rule or order if he determines on the record, after opportunity for
an inspection where appropriate and a hearing, that the proponent of
the variance has demonstrated by a preponderance of the evidence that
the conditions, practices, means, methods, operations, or processes
used or proposed to be used by an employer will provide employment
and places of employment to his employees which are as safe and
healthful as those which would prevail if he complied with the
standard. The rule or order so issued shall prescribe the conditions
the employer must maintain, and the practices, means, methods, operations, and processes which he must adopt and utilize to the extent they

Pub. Law 91-596 — 8 — **December 29, 1970**

differ from the standard in question. Such a rule or order may be modified or revoked upon application by an employer, employees, or by the Secretary on his own motion, in the manner prescribed for its issuance under this subsection at any time after six months from its issuance.

Publication
in Federal
Register.

(e) Whenever the Secretary promulgates any standard, makes any rule, order, or decision, grants any exemption or extension of time, or compromises, mitigates, or settles any penalty assessed under this Act, he shall include a statement of the reasons for such action, which shall be published in the Federal Register.

Petition for
Judicial
review.

(f) Any person who may be adversely affected by a standard issued under this section may at any time prior to the sixtieth day after such standard is promulgated file a petition challenging the validity of such standard with the United States court of appeals for the circuit wherein such person resides or has his principal place of business, for a judicial review of such standard. A copy of the petition shall be forthwith transmitted by the clerk of the court to the Secretary. The filing of such petition shall not, unless otherwise ordered by the court, operate as a stay of the standard. The determinations of the Secretary shall be conclusive if supported by substantial evidence in the record considered as a whole.

(g) In determining the priority for establishing standards under this section, the Secretary shall give due regard to the urgency of the need for mandatory safety and health standards for particular industries, trades, crafts, occupations, businesses, workplaces or work environments. The Secretary shall also give due regard to the recommendations of the Secretary of Health, Education, and Welfare regarding the need for mandatory standards in determining the priority for establishing such standards.

ADVISORY COMMITTEES; ADMINISTRATION

Establishment;
membership.

Sec. 7. (a)(1) There is hereby established a National Advisory Committee on Occupational Safety and Health consisting of twelve members appointed by the Secretary, four of whom are to be designated by the Secretary of Health, Education, and Welfare, without

80 Stat. 378.
5 USC 101.

regard to the provisions of title 5, United States Code, governing appointments in the competitive service, and composed of representatives of management, labor, occupational safety and occupational health professions, and of the public. The Secretary shall designate one of the public members as Chairman. The members shall be selected upon the basis of their experience and competence in the field of occupational safety and health.

(2) The Committee shall advise, consult with, and make recommendations to the Secretary and the Secretary of Health, Education, and Welfare on matters relating to the administration of the Act. The Committee shall hold no fewer than two meetings during each calen-

Public tran-
script.

dar year. All meetings of the Committee shall be open to the public and a transcript shall be kept and made available for public inspection.

(3) The members of the Committee shall be compensated in accordance with the provisions of section 3109 of title 5, United States

80 Stat. 416.

Code.

(4) The Secretary shall furnish to the Committee an executive secretary and such secretarial, clerical, and other services as are deemed necessary to the conduct of its business.

(b) An advisory committee may be appointed by the Secretary to assist him in his standard-setting functions under section 6 of this Act. Each such committee shall consist of not more than fifteen members

December 29, 1970   **- 9 -**   **Pub. Law 91-596**   84 STAT. 1598

and shall include as a member one or more designees of the Secretary of Health, Education, and Welfare, and shall include among its members an equal number of persons qualified by experience and affiliation to present the viewpoint of the employers involved, and of persons similarly qualified to present the viewpoint of the workers involved, as well as one or more represent..tives of health and safety agencies of the States. An advisory committee may also include such other persons as the Secretary may appoint who are qualified by knowledge and experience to make a useful contribution to the work of such committee, including one or more representatives of professional organizations of technicians or professionals specializing in occupational safety or health, and one or more representatives of nationally recognized standards-producing organizations, but the number of persons so appointed to any such advisory committee shall not exceed the number appointed to such committee as representatives of Federal and State agencies. Persons appointed to advisory committees from private life shall be compensated in the same manner as consultants or experts under section 3109 of title 5, United States Code. The Secre-   80 Stat. 416. tary shall pay to any State which is the employer of a member of such a committee who is a representative of the health or safety agency of that State, reimbursement sufficient to cover the actual cost to the State resulting from such representative's membership on such committee. Any meeting of such committee shall be open to the public   Recordkeeping. and an accurate record shall be kept and made available to the public. No member of such committee (other than representatives of employers and employees) shall have an economic interest in any proposed rule.

(c) In carrying out his responsibilities under this Act, the Secretary is authorized to—

(1) use, with the consent of any Federal agency, the services, facilities, and personnel of such agency, with or without reimbursement, and with the consent of any State or political subdivision thereof, accept and use the services, facilities, and personnel of any agency of such State or subdivision with reimbursement; and

(2) employ experts and consultants or organizations thereof as authorized by section 3109 of title 5, United States Code, except that contracts for such employment may be renewed annually; compensate individuals so employed at rates not in excess of the rate specified at the time of service for grade GS-18 under section 5332 of title 5, United States Code, including traveltime, and   Ante, p. 198-1. allow them while away from their homes or regular places of business, travel expenses (including per diem in lieu of subsistence) as authorized by section 5703 of title 5, United States Code, for per-   80 Stat. 499; sons in the Government service employed intermittently, while so   83 Stat. 190. employed.

INSPECTIONS. INVESTIGATIONS. AND RECORDKEEPING

SEC. 8. (a) In order to carry out the purposes of this Act, the Secretary, upon presenting appropriate credentials to the owner, operator, or agent in charge, is authorized—

(1) to enter without delay and at reasonable times any factory, plant, establishment, construction site, or other area, workplace or environment where work is performed by an employee of an employer; and

84 STAT. 1599

(2) to inspect and investigate during regular working hours and at other reasonable times, and within reasonable limits and in a reasonable manner, any such place of employment and all pertinent conditions, structures, machines, apparatus, devices, equipment, and materials therein, and to question privately any such employer, owner, operator, agent or employee.

Subpoena power.

(b) In making his inspections and investigations under this Act the Secretary may require the attendance and testimony of witnesses and the production of evidence under oath. Witnesses shall be paid the same fees and mileage that are paid witnesses in the courts of the United States. In case of a contumacy, failure, or refusal of any person to obey such an order, any district court of the United States or the United States courts of any territory or possession, within the jurisdiction of which such person is found, or resides or transacts business, upon the application by the Secretary, shall have jurisdiction to issue to such person an order requiring such person to appear to produce evidence if, as, and when so ordered, and to give testimony relating to the matter under investigation or in question, and any failure to obey such order of the court may be punished by said court as a contempt thereof.

Recordkeeping.

(c) (1) Each employer shall make, keep and preserve, and make available to the Secretary or the Secretary of Health, Education, and Welfare, such records regarding his activities relating to this Act as the Secretary, in cooperation with the Secretary of Health, Education, and Welfare, may prescribe by regulation as necessary or appropriate for the enforcement of this Act or for developing information regarding the causes and prevention of occupational accidents and illnesses. In order to carry out the provisions of this paragraph such regulations may include provisions requiring employers to conduct periodic inspections. The Secretary shall also issue regulations requiring that employers, through posting of notices or other appropriate means, keep their employees informed of their protections and obligations under this Act, including the provisions of applicable standards.

Work-related deaths, etc.; reports.

(2) The Secretary, in cooperation with the Secretary of Health, Education, and Welfare, shall prescribe regulations requiring employers to maintain accurate records of, and to make periodic reports on, work-related deaths, injuries and illnesses other than minor injuries requiring only first aid treatment and which do not involve medical treatment, loss of consciousness, restriction of work or motion, or transfer to another job.

(3) The Secretary, in cooperation with the Secretary of Health, Education, and Welfare, shall issue regulations requiring employers to maintain accurate records of employee exposures to potentially toxic materials or harmful physical agents which are required to be monitored or measured under section 6. Such regulations shall provide employees or their representatives with an opportunity to observe such monitoring or measuring, and to have access to the records thereof. Such regulations shall also make appropriate provision for each employee or former employee to have access to such records as will indicate his own exposure to toxic materials or harmful physical agents. Each employer shall promptly notify any employee who has been or is being exposed to toxic materials or harmful physical agents in concentrations or at levels which exceed those prescribed by an applicable occupational safety and health standard promulgated under section 6, and shall inform any employee who is being thus exposed of the corrective action being taken.

December 29, 1970   - 11 -   Pub. Law 91-596
84 STAT. 1600

(d) Any information obtained by the Secretary, the Secretary of Health, Education, and Welfare, or a State agency under this Act shall be obtained with a minimum burden upon employers, especially those operating small businesses. Unnecessary duplication of efforts in obtaining information shall be reduced to the maximum extent feasible.

(e) Subject to regulations issued by the Secretary, a representative of the employer and a representative authorized by his employees shall be given an opportunity to accompany the Secretary or his authorized representative during the physical inspection of any workplace under subsection (a) for the purpose of aiding such inspection. Where there is no authorized employee representative, the Secretary or his authorized representative shall consult with a reasonable number of employees concerning matters of health and safety in the workplace.

(f)(1) Any employees or representative of employees who believe that a violation of a safety or health standard exists that threatens physical harm, or that an imminent danger exists, may request an inspection by giving notice to the Secretary or his authorized representative of such violation or danger. Any such notice shall be reduced to writing, shall set forth with reasonable particularity the grounds for the notice, and shall be signed by the employees or representative of employees, and a copy shall be provided the employer or his agent no later than at the time of inspection, except that, upon the request of the person giving such notice, his name and the names of individual employees referred to therein shall not appear in such copy or on any record published, released, or made available pursuant to subsection (g) of this section. If upon receipt of such notification the Secretary determines there are reasonable grounds to believe that such violation or danger exists, he shall make a special inspection in accordance with the provisions of this section as soon as practicable, to determine if such violation or danger exists. If the Secretary determines there are no reasonable grounds to believe that a violation or danger exists he shall notify the employees or representative of the employees in writing of such determination.

(2) Prior to or during any inspection of a workplace, any employees or representative of employees employed in such workplace may notify the Secretary or any representative of the Secretary responsible for conducting the inspection, in writing, of any violation of this Act which they have reason to believe exists in such workplace. The Secretary shall, by regulation, establish procedures for informal review of any refusal by a representative of the Secretary to issue a citation with respect to any such alleged violation and shall furnish the employees or representative of employees requesting such review a written statement of the reasons for the Secretary's final disposition of the case.

(g)(1) The Secretary and Secretary of Health, Education, and Welfare are authorized to compile, analyze, and publish, either in summary or detailed form, all reports or information obtained under this section. **Reports, publication.**

(2) The Secretary and the Secretary of Health, Education, and Welfare shall each prescribe such rules and regulations as he may deem necessary to carry out their responsibilities under this Act, including rules and regulations dealing with the inspection of an employer's establishment. **Rules and regulations.**

84 STAT. 1601

CITATIONS

Sec. 9. (a) If, upon inspection or investigation, the Secretary or his authorized representative believes that an employer has violated a requirement of section 5 of this Act, of any standard, rule or order promulgated pursuant to section 6 of this Act, or of any regulations prescribed pursuant to this Act, he shall with reasonable promptness issue a citation to the employer. Each citation shall be in writing and shall describe with particularity the nature of the violation, including a reference to the provision of the Act, standard, rule, regulation, or order alleged to have been violated. In addition, the citation shall fix a reasonable time for the abatement of the violation. The Secretary may prescribe procedures for the issuance of a notice in lieu of a citation with respect to de minimis violations which have no direct or immediate relationship to safety or health.

(b) Each citation issued under this section, or a copy or copies thereof, shall be prominently posted, as prescribed in regulations issued by the Secretary, at or near each place a violation referred to in the citation occurred.

Limitation.

(c) No citation may be issued under this section after the expiration of six months following the occurrence of any violation.

PROCEDURE FOR ENFORCEMENT

Sec. 10. (a) If, after an inspection or investigation, the Secretary issues a citation under section 9(a), he shall, within a reasonable time after the termination of such inspection or investigation, notify the employer by certified mail of the penalty, if any, proposed to be assessed under section 17 and that the employer has fifteen working days within which to notify the Secretary that he wishes to contest the citation or proposed assessment of penalty. If, within fifteen working days from the receipt of the notice issued by the Secretary the employer fails to notify the Secretary that he intends to contest the citation or proposed assessment of penalty, and no notice is filed by any employee or representative of employees under subsection (c) within such time, the citation and the assessment, as proposed, shall be deemed a final order of the Commission and not subject to review by any court or agency.

(b) If the Secretary has reason to believe that an employer has failed to correct a violation for which a citation has been issued within the period permitted for its correction (which period shall not begin to run until the entry of a final order by the Commission in the case of any review proceedings under this section initiated by the employer in good faith and not solely for delay or avoidance of penalties), the Secretary shall notify the employer by certified mail of such failure and of the penalty proposed to be assessed under section 17 by reason of such failure, and that the employer has fifteen working days within which to notify the Secretary that he wishes to contest the Secretary's notification or the proposed assessment of penalty. If, within fifteen working days from the receipt of notification issued by the Secretary, the employer fails to notify the Secretary that he intends to contest the notification or proposed assessment of penalty, the notification and assessment, as proposed, shall be deemed a final order of the Commission and not subject to review by any court or agency.

(c) If an employer notifies the Secretary that he intends to contest a citation issued under section 9(a) or notification issued under subsection (a) or (b) of this section, or if, within fifteen working days

84 STAT. 1602

of the issuance of a citation under section 9(a), any employee or representative of employees files a notice with the Secretary alleging that the period of time fixed in the citation for the abatement of the violation is unreasonable, the Secretary shall immediately advise the Commission of such notification, and the Commission shall afford an opportunity for a hearing (in accordance with section 554 of title 5, United States Code, but without regard to subsection (a)(3) of such 80 Stat. 384. section). The Commission shall thereafter issue an order, based on findings of fact, affirming, modifying, or vacating the Secretary's citation or proposed penalty, or directing other appropriate relief, and such order shall become final thirty days after its issuance. Upon a showing by an employer of a good faith effort to comply with the abatement requirements of a citation, and that abatement has not been completed because of factors beyond his reasonable control, the Secretary, after an opportunity for a hearing as provided in this subsection, shall issue an order affirming or modifying the abatement requirements in such citation. The rules of procedure prescribed by the Commission shall provide affected employees or representatives of affected employees an opportunity to participate as parties to hearings under this subsection.

JUDICIAL REVIEW

SEC. 11. (a) Any person adversely affected or aggrieved by an order of the Commission issued under subsection (c) of section 10 may obtain a review of such order in any United States court of appeals for the circuit in which the violation is alleged to have occurred or where the employer has its principal office, or in the Court of Appeals for the District of Columbia Circuit, by filing in such court within sixty days following the issuance of such order a written petition praying that the order be modified or set aside. A copy of such petition shall be forthwith transmitted by the clerk of the court to the Commission and to the other parties, and thereupon the Commission shall file in the court the record in the proceeding as provided in section 2112 of title 28, United States Code. Upon such filing, the court shall have jurisdiction 72 Stat. 941; of the proceeding and of the question determined therein, and shall 80 Stat. 1323. have power to grant such temporary relief or restraining order as it deems just and proper, and to make and enter upon the pleadings, testimony, and proceedings set forth in such record a decree affirming, modifying, or setting aside in whole or in part, the order of the Commission and enforcing the same to the extent that such order is affirmed or modified. The commencement of proceedings under this subsection shall not, unless ordered by the court, operate as a stay of the order of the Commission. No objection that has not been urged before the Commission shall be considered by the court, unless the failure or neglect to urge such objection shall be excused because of extraordinary circumstances. The findings of the Commission with respect to questions of fact, if supported by substantial evidence on the record considered as a whole, shall be conclusive. If any party shall apply to the court for leave to adduce additional evidence and shall show to the satisfaction of the court that such additional evidence is material and that there were reasonable grounds for the failure to adduce such evidence in the hearing before the Commission, the court may order such additional evidence to be taken before the Commission and to be made a part of the record. The Commission may modify its findings as to the facts, or make new findings, by reason of additional evidence so taken and filed, and it shall file such modified or new findings, which findings with respect to questions of fact, if supported by substantial evi-

Pub. Law 91-596   - 14 -   **December 29, 1970**

84 STAT. 1603

dence on the record considered as a whole, shall be conclusive, and its recommendations, if any, for the modification or setting aside of its original order. Upon the filing of the record with it, the jurisdiction of the court shall be exclusive and its judgment and decree shall be final, except that the same shall be subject to review by the Supreme Court of the United States, as provided in section 1254 of title 28, United States Code. Petitions filed under this subsection shall be heard expeditiously.

62 Stat. 928.

(b) The Secretary may also obtain review or enforcement of any final order of the Commission by filing a petition for such relief in the United States court of appeals for the circuit in which the alleged violation occurred or in which the employer has its principal office, and the provisions of subsection (a) shall govern such proceedings to the extent applicable. If no petition for review, as provided in subsection (a), is filed within sixty days after service of the Commission's order, the Commission's findings of fact and order shall be conclusive in connection with any petition for enforcement which is filed by the Secretary after the expiration of such sixty-day period. In any such case, as well as in the case of a noncontested citation or notification by the Secretary which has become a final order of the Commission under subsection (a) or (b) of section 10, the clerk of the court, unless otherwise ordered by the court, shall forthwith enter a decree enforcing the order and shall transmit a copy of such decree to the Secretary and the employer named in the petition. In any contempt proceeding brought to enforce a decree of a court of appeals entered pursuant to this subsection or subsection (a), the court of appeals may assess the penalties provided in section 17, in addition to invoking any other available remedies.

(c) (1) No person shall discharge or in any manner discriminate against any employee because such employee has filed any complaint or instituted or caused to be instituted any proceeding under or related to this Act or has testified or is about to testify in any such proceeding or because of the exercise by such employee on behalf of himself or others of any right afforded by this Act.

(2) Any employee who believes that he has been discharged or otherwise discriminated against by any person in violation of this subsection may, within thirty days after such violation occurs, file a complaint with the Secretary alleging such discrimination. Upon receipt of such complaint, the Secretary shall cause such investigation to be made as he deems appropriate. If upon such investigation, the Secretary determines that the provisions of this subsection have been violated, he shall bring an action in any appropriate United States district court against such person. In any such action the United States district courts shall have jurisdiction, for cause shown to restrain violations of paragraph (1) of this subsection and order all appropriate relief including rehiring or reinstatement of the employee to his former position with back pay.

(3) Within 90 days of the receipt of a complaint filed under the subsection the Secretary shall notify the complainant of his determination under paragraph 2 of this subsection.

THE OCCUPATIONAL SAFETY AND HEALTH REVIEW COMMISSION

Establishment; membership.

SEC. 12. (a) The Occupational Safety and Health Review Commission is hereby established. The Commission shall be composed of three members who shall be appointed by the President, by and with the advice and consent of the Senate, from among persons who by reason

84 STAT. 1604

of training, education, or experience are qualified to carry out the
functions of the Commission under this Act. The President shall designate one of the members of the Commission to serve as Chairman.

(b) The terms of members of the Commission shall be six years   Terms.
except that (1) the members of the Commission first taking office shall
serve, as designated by the President at the time of appointment, one
for a term of two years, one for a term of four years, and one for a
term of six years, and (2) a vacancy caused by the death, resignation,
or removal of a member prior to the expiration of the term for which
he was appointed shall be filled only for the remainder of such
unexpired term. A member of the Commission may be removed by the
President for inefficiency, neglect of duty, or malfeasance in office.

(c) (1) Section 5314 of title 5, United States Code, is amended by   80 Stat. 460.
adding at the end thereof the following new paragraph:
    "(57) Chairman, Occupational Safety and Health Review
Commission."

(2) Section 5315 of title 5, United States Code, is amended by add-   Ante, p. 776.
ing at the end thereof the following new paragraph:
    "(94) Members, Occupational Safety and Health Review
Commission."

(d) The principal office of the Commission shall be in the District   Location.
of Columbia. Whenever the Commission deems that the convenience
of the public or of the parties may be promoted, or delay or expense
may be minimized, it may hold hearings or conduct other proceedings
at any other place.

(e) The Chairman shall be responsible on behalf of the Commission
for the administrative operations of the Commission and shall appoint
such hearing examiners and other employees as he deems necessary
to assist in the performance of the Commission's functions and to
fix their compensation in accordance with the provisions of chapter
51 and subchapter III of chapter 53 of title 5, United States Code,   5 USC 5101,
relating to classification and General Schedule pay rates: *Provided,*   5331.
That assignment, removal and compensation of hearing examiners   Ante, p. 198-1.
shall be in accordance with sections 3105, 3344, 5362, and 7521 of title 5,
United States Code.

(f) For the purpose of carrying out its functions under this Act, two   Quorum.
members of the Commission shall constitute a quorum and official
action can be taken only on the affirmative vote of at least two
members.

(g) Every official act of the Commission shall be entered of record,   Public records.
and its hearings and records shall be open to the public. The Commission is authorized to make such rules as are necessary for the orderly
transaction of its proceedings. Unless the Commission has adopted a
different rule, its proceedings shall be in accordance with the Federal
Rules of Civil Procedure.   28 USC app.

(h) The Commission may order testimony to be taken by deposition
in any proceedings pending before it at any state of such proceeding.
Any person may be compelled to appear and depose, and to produce
books, papers, or documents, in the same manner as witnesses may be
compelled to appear and testify and produce like documentary
evidence before the Commission. Witnesses whose depositions are taken
under this subsection, and the persons taking such depositions, shall be
entitled to the same fees as are paid for like services in the courts of
the United States.

(i) For the purpose of any proceeding before the Commission, the
provisions of section 11 of the National Labor Relations Act (29
U.S.C. 161) are hereby made applicable to the jurisdiction and powers   61 Stat. 150;
of the Commission.   Ante, p. 930.

84 STAT. 1605

Report.

(j) A hearing examiner appointed by the Commission shall hear, and make a determination upon, any proceeding instituted before the Commission and any motion in connection therewith, assigned to such hearing examiner by the Chairman of the Commission, and shall make a report of any such determination which constitutes his final disposition of the proceedings. The report of the hearing examiner shall become the final order of the Commission within thirty days after such report by the hearing examiner, unless within such period any Commission member has directed that such report shall be reviewed by the Commission.

(k) Except as otherwise provided in this Act, the hearing examiners shall be subject to the laws governing employees in the classified civil service, except that appointments shall be made without regard to

80 Stat. 453.

section 5108 of title 5, United States Code. Each hearing examiner shall receive compensation at a rate not less than that prescribed for

Ante, p. 198-1.

GS-16 under section 5332 of title 5, United States Code.

PROCEDURES TO COUNTERACT IMMINENT DANGERS

SEC. 13. (a) The United States district courts shall have jurisdiction, upon petition of the Secretary, to restrain any conditions or practices in any place of employment which are such that a danger exists which could reasonably be expected to cause death or serious physical harm immediately or before the imminence of such danger can be eliminated through the enforcement procedures otherwise provided by this Act. Any order issued under this section may require such steps to be taken as may be necessary to avoid, correct, or remove such imminent danger and prohibit the employment or presence of any individual in locations or under conditions where such imminent danger exists, except individuals whose presence is necessary to avoid, correct, or remove such imminent danger or to maintain the capacity of a continuous process operation to resume normal operations without a complete cessation of operations, or where a cessation of operations is necessary, to permit such to be accomplished in a safe and orderly manner.

(b) Upon the filing of any such petition the district court shall have jurisdiction to grant such injunctive relief or temporary restraining order pending the outcome of an enforcement proceeding pursuant to this Act. The proceeding shall be as provided by Rule 65 of the Fed-

28 USC app.

eral Rules, Civil Procedure, except that no temporary restraining order issued without notice shall be effective for a period longer than five days.

(c) Whenever and as soon as an inspector concludes that conditions or practices described in subsection (a) exist in any place of employment, he shall inform the affected employees and employers of the danger and that he is recommending to the Secretary that relief be sought.

(d) If the Secretary arbitrarily or capriciously fails to seek relief under this section, any employee who may be injured by reason of such failure, or the representative of such employees, might bring an action against the Secretary in the United States district court for the district in which the imminent danger is alleged to exist or the employer has its principal office, or for the District of Columbia, for a writ of mandamus to compel the Secretary to seek such an order and for such further relief as may be appropriate.

84 STAT. 1606

REPRESENTATION IN CIVIL LITIGATION

SEC. 14. Except as provided in section 518(a) of title 28, United States Code, relating to litigation before the Supreme Court, the 80 Stat. 613. Solicitor of Labor may appear for and represent the Secretary in any civil litigation brought under this Act but all such litigation shall be subject to the direction and control of the Attorney General.

CONFIDENTIALITY OF TRADE SECRETS

SEC. 15. All information reported to or otherwise obtained by the Secretary or his representative in connection with any inspection or proceeding under this Act which contains or which might reveal a trade secret referred to in section 1905 of title 18 of the United States Code shall be considered confidential for the purpose of that section, 62 Stat. 791. except that such information may be disclosed to other officers or employees concerned with carrying out this Act or when relevant in any proceeding under this Act. In any such proceeding the Secretary, the Commission, or the court shall issue such orders as may be appropriate to protect the confidentiality of trade secrets.

VARIATIONS, TOLERANCES, AND EXEMPTIONS

SEC. 16. The Secretary, on the record, after notice and opportunity for a hearing may provide such reasonable limitations and may make such rules and regulations allowing reasonable variations, tolerances, and exemptions to and from any or all provisions of this Act as he may find necessary and proper to avoid serious impairment of the national defense. Such action shall not be in effect for more than six months without notification to affected employees and an opportunity being afforded for a hearing.

PENALTIES

SEC. 17. (a) Any employer who willfully or repeatedly violates the requirements of section 5 of this Act, any standard, rule, or order promulgated pursuant to section 6 of this Act, or regulations prescribed pursuant to this Act, may be assessed a civil penalty of not more than $10,000 for each violation.

(b) Any employer who has received a citation for a serious violation of the requirements of section 5 of this Act, of any standard, rule, or order promulgated pursuant to section 6 of this Act, or of any regulations prescribed pursuant to this Act, shall be assessed a civil penalty of up to $1,000 for each such violation.

(c) Any employer who has received a citation for a violation of the requirements of section 5 of this Act, of any standard, rule, or order promulgated pursuant to section 6 of this Act, or of regulations prescribed pursuant to this Act, and such violation is specifically determined not to be of a serious nature, may be assessed a civil penalty of up to $1,000 for each such violation.

(d) Any employer who fails to correct a violation for which a citation has been issued under section 9(a) within the period permitted for its correction (which period shall not begin to run until the date of the final order of the Commission in the case of any review proceeding under section 10 initiated by the employer in good faith and not solely for delay or avoidance of penalties), may be assessed a civil penalty of not more than $1,000 for each day during which such failure or violation continues.

84 STAT. 1607

(e) Any employer who willfully violates any standard, rule, or order promulgated pursuant to section 6 of this Act, or of any regulations prescribed pursuant to this Act, and that violation caused death to any employee, shall, upon conviction, be punished by a fine of not more than $10,000 or by imprisonment for not more than six months, or by both; except that if the conviction is for a violation committed after a first conviction of such person, punishment shall be by a fine of not more than $20,000 or by imprisonment for not more than one year, or by both.

(f) Any person who gives advance notice of any inspection to be conducted under this Act, without authority from the Secretary or his designees, shall, upon conviction, be punished by a fine of not more than $1,000 or by imprisonment for not more than six months, or by both.

(g) Whoever knowingly makes any false statement, representation, or certification in any application, record, report, plan, or other document filed or required to be maintained pursuant to this Act shall, upon conviction, be punished by a fine of not more than $10,000, or by imprisonment for not more than six months, or by both.

65 Stat. 721;
79 Stat. 234.

(h)(1) Section 1114 of title 18, United States Code, is hereby amended by striking out "designated by the Secretary of Health, Education, and Welfare to conduct investigations, or inspections under the Federal Food, Drug, and Cosmetic Act" and inserting in lieu thereof "or of the Department of Labor assigned to perform investigative, inspection, or law enforcement functions".

62 Stat. 756.

(2) Notwithstanding the provisions of sections 1111 and 1114 of title 18, United States Code, whoever, in violation of the provisions of section 1114 of such title, kills a person while engaged in or on account of the performance of investigative, inspection, or law enforcement functions added to such section 1114 by paragraph (1) of this subsection, and who would otherwise be subject to the penalty provisions of such section 1111, shall be punished by imprisonment for any term of years or for life.

(i) Any employer who violates any of the posting requirements, as prescribed under the provisions of this Act, shall be assessed a civil penalty of up to $1,000 for each violation.

(j) The Commission shall have authority to assess all civil penalties provided in this section, giving due consideration to the appropriateness of the penalty with respect to the size of the business of the employer being charged, the gravity of the violation, the good faith of the employer, and the history of previous violations.

(k) For purposes of this section, a serious violation shall be deemed to exist in a place of employment if there is a substantial probability that death or serious physical harm could result from a condition which exists, or from one or more practices, means, methods, operations, or processes which have been adopted or are in use, in such place of employment unless the employer did not, and could not with the exercise of reasonable diligence, know of the presence of the violation.

(l) Civil penalties owed under this Act shall be paid to the Secretary for deposit into the Treasury of the United States and shall accrue to the United States and may be recovered in a civil action in the name of the United States brought in the United States district court for the district where the violation is alleged to have occurred or where the employer has its principal office.

STATE JURISDICTION AND STATE PLANS

SEC. 18. (a) Nothing in this Act shall prevent any State agency or court from asserting jurisdiction under State law over any occupational safety or health issue with respect to which no standard is in effect under section 6.

(b) Any State which, at any time, desires to assume responsibility for development and enforcement therein of occupational safety and health standards relating to any occupational safety or health issue with respect to which a Federal standard has been promulgated under section 6 shall submit a State plan for the development of such standards and their enforcement.

(c) The Secretary shall approve the plan submitted by a State under subsection (b), or any modification thereof, if such plan in his judgment—

(1) designates a State agency or agencies as the agency or agencies responsible for administering the plan throughout the State,

(2) provides for the development and enforcement of safety and health standards relating to one or more safety or health issues, which standards (and the enforcement of which standards) are or will be at least as effective in providing safe and healthful employment and places of employment as the standards promulgated under section 6 which relate to the same issues, and which standards, when applicable to products which are distributed or used in interstate commerce, are required by compelling local conditions and do not unduly burden interstate commerce,

(3) provides for a right of entry and inspection of all workplaces subject to the Act which is at least as effective as that provided in section 8, and includes a prohibition on advance notice of inspections,

(4) contains satisfactory assurances that such agency or agencies have or will have the legal authority and qualified personnel necessary for the enforcement of such standards,

(5) gives satisfactory assurances that such State will devote adequate funds to the administration and enforcement of such standards,

(6) contains satisfactory assurances that such State will, to the extent permitted by its law, establish and maintain an effective and comprehensive occupational safety and health program applicable to all employees of public agencies of the State and its political subdivisions, which program is as effective as the standards contained in an approved plan,

(7) requires employers in the State to make reports to the Secretary in the same manner and to the same extent as if the plan were not in effect, and

(8) provides that the State agency will make such reports to the Secretary in such form and containing such information, as the Secretary shall from time to time require.

(d) If the Secretary rejects a plan submitted under subsection (b), he shall afford the State submitting the plan due notice and opportunity for a hearing before so doing. *Notice of hearing.*

(e) After the Secretary approves a State plan submitted under subsection (b), he may, but shall not be required to, exercise his authority under sections 8, 9, 10, 13, and 17 with respect to comparable standards promulgated under section 6, for the period specified in the next sentence. The Secretary may exercise the authority referred to above until he determines, on the basis of actual operations under the

84 STAT. 1609

State plan, that the criteria set forth in subsection (c) are being applied, but he shall not make such determination for at least three years after the plan's approval under subsection (c). Upon making the determination referred to in the preceding sentence, the provisions of sections 5(a)(2), 8 (except for the purpose of carrying out subsection (f) of this section), 9, 10, 13, and 17, and standards promulgated under section 6 of this Act, shall not apply with respect to any occupational safety or health issues covered under the plan, but the Secretary may retain jurisdiction under the above provisions in any proceeding commenced under section 9 or 10 before the date of determination.

Continuing evaluation.

(f) The Secretary shall, on the basis of reports submitted by the State agency and his own inspections make a continuing evaluation of the manner in which each State having a plan approved under this section is carrying out such plan. Whenever the Secretary finds, after affording due notice and opportunity for a hearing, that in the administration of the State plan there is a failure to comply substantially with any provision of the State plan (or any assurance contained therein), he shall notify the State agency of his withdrawal of approval of such plan and upon receipt of such notice such plan shall cease to be in effect, but the State may retain jurisdiction in any case commenced before the withdrawal of the plan in order to enforce standards under the plan whenever the issues involved do not relate to the reasons for the withdrawal of the plan.

Plan rejection, review.

(g) The State may obtain a review of a decision of the Secretary withdrawing approval of or rejecting its plan by the United States court of appeals for the circuit in which the State is located by filing in such court within thirty days following receipt of notice of such decision a petition to modify or set aside in whole or in part the action of the Secretary. A copy of such petition shall forthwith be served upon the Secretary, and thereupon the Secretary shall certify and file in the court the record upon which the decision complained of was issued as provided in section 2112 of title 28, United States Code. Unless the court finds that the Secretary's decision in rejecting a proposed State plan or withdrawing his approval of such a plan is not supported by substantial evidence the court shall affirm the Secretary's decision. The judgment of the court shall be subject to review by the Supreme Court of the United States upon certiorari or certification as provided in section 1254 of title 28, United States Code.

72 Stat. 941;
80 Stat. 1323.

62 Stat. 928.

(h) The Secretary may enter into an agreement with a State under which the State will be permitted to continue to enforce one or more occupational health and safety standards in effect in such State until final action is taken by the Secretary with respect to a plan submitted by a State under subsection (b) of this section, or two years from the date of enactment of this Act, whichever is earlier.

FEDERAL AGENCY SAFETY PROGRAMS AND RESPONSIBILITIES

SEC. 19. (a) It shall be the responsibility of the head of each Federal agency to establish and maintain an effective and comprehensive occupational safety and health program which is consistent with the standards promulgated under section 6. The head of each agency shall (after consultation with representatives of the employees thereof)—

(1) provide safe and healthful places and conditions of employment, consistent with the standards set under section 6;

(2) acquire, maintain, and require the use of safety equipment, personal protective equipment, and devices reasonably necessary to protect employees;

208   ACCIDENT PREVENTION AND OSHA COMPLIANCE

December 29, 1970   - 21 -   Pub. Law 91-596

(3) keep adequate records of all occupational accidents and ill-

nesses for proper evaluation and necessary corrective action; Recordkeeping.

(4) consult with the Secretary with regard to the adequacy as
to form and content of records kept pursuant to subsection (a)(3)
of this section; and

(5) make an annual report to the Secretary with respect to Annual report.
occupational accidents and injuries and the agency's program
under this section. Such report shall include any report submitted
under section 7902(e)(2) of title 5, United States Code. 80 Stat. 530.

(b) The Secretary shall report to the President a summary or digest Report to
of reports submitted to him under subsection (a)(5) of this section, President.
together with his evaluations of and recommendations derived from
such reports. The President shall transmit annually to the Senate and Report to
the House of Representatives a report of the activities of Federal Congress.
agencies under this section.

(c) Section 7902(c)(1) of title 5, United States Code, is amended
by inserting after "agencies" the following: "and of labor organizations representing employees".

(d) The Secretary shall have access to records and reports kept Records, etc.;
and filed by Federal agencies pursuant to subsections (a)(3) and (5) availability.
of this section unless those records and reports are specifically required
by Executive order to be kept secret in the interest of the national
defense or foreign policy, in which case the Secretary shall have access
to such information as will not jeopardize national defense or foreign
policy.

RESEARCH AND RELATED ACTIVITIES

SEC. 20. (a)(1) The Secretary of Health, Education, and Welfare,
after consultation with the Secretary and with other appropriate
Federal departments or agencies, shall conduct (directly or by grants
or contracts) research, experiments, and demonstrations relating to
occupational safety and health, including studies of psychological
factors involved, and relating to innovative methods, techniques, and
approaches for dealing with occupational safety and health problems.

(2) The Secretary of Health, Education, and Welfare shall from
time to time consult with the Secretary in order to develop specific
plans for such research, demonstrations, and experiments as are necessary to produce criteria, including criteria identifying toxic substances, enabling the Secretary to meet his responsibility for the
formulation of safety and health standards under this Act; and the
Secretary of Health, Education, and Welfare, on the basis of such
research, demonstrations, and experiments and any other information
available to him, shall develop and publish at least annually such
criteria as will effectuate the purposes of this Act.

(3) The Secretary of Health, Education, and Welfare, on the basis
of such research, demonstrations, and experiments, and any other
information available to him, shall develop criteria dealing with toxic
materials and harmful physical agents and substances which will
describe exposure levels that are safe for various periods of employment, including but not limited to the exposure levels at which no
employee will suffer impaired health or functional capacities or
diminished life expectancy as a result of his work experience.

(4) The Secretary of Health, Education, and Welfare shall also
conduct special research, experiments, and demonstrations relating
to occupational safety and health as are necessary to explore new
problems, including those created by new technology in occupational
safety and health, which may require ameliorative action beyond that

84 STAT. 1611

which is otherwise provided for in the operating provisions of this Act. The Secretary of Health, Education, and Welfare shall also conduct research into the motivational and behavioral factors relating to the field of occupational safety and health.

Toxic substances, records.

(5) The Secretary of Health, Education, and Welfare, in order to comply with his responsibilities under paragraph (2), and in order to develop needed information regarding potentially toxic substances or harmful physical agents, may prescribe regulations requiring employers to measure, record, and make reports on the exposure of employees to substances or physical agents which the Secretary of Health, Education, and Welfare reasonably believes may endanger the health or safety of employees. The Secretary of Health, Education,

Medical examinations.

and Welfare also is authorized to establish such programs of medical examinations and tests as may be necessary for determining the incidence of occupational illnesses and the susceptibility of employees to such illnesses. Nothing in this or any other provision of this Act shall be deemed to authorize or require medical examination, immunization, or treatment for those who object thereto on religious grounds, except where such is necessary for the protection of the health or safety of others. Upon the request of any employer who is required to measure and record exposure of employees to substances or physical agents as provided under this subsection, the Secretary of Health, Education, and Welfare shall furnish full financial or other assistance to such employer for the purpose of defraying any additional expense incurred by him in carrying out the measuring and recording as provided in this subsection.

Toxic substances, publication.

(6) The Secretary of Health, Education, and Welfare shall publish within six months of enactment of this Act and thereafter as needed but at least annually a list of all known toxic substances by generic family or other useful grouping, and the concentrations at which such toxicity is known to occur. He shall determine following a written request by any employer or authorized representative of employees, specifying with reasonable particularity the grounds on which the request is made, whether any substance normally found in the place of employment has potentially toxic effects in such concentrations as used or found; and shall submit such determination both to employers and affected employees as soon as possible. If the Secretary of Health, Education, and Welfare determines that any substance is potentially toxic at the concentrations in which it is used or found in a place of employment, and such substance is not covered by an occupational safety or health standard promulgated under section 6, the Secretary of Health, Education, and Welfare shall immediately submit such determination to the Secretary, together with all pertinent criteria.

Annual studies.

(7) Within two years of enactment of this Act, and annually thereafter the Secretary of Health, Education, and Welfare shall conduct and publish industrywide studies of the effect of chronic or low-level exposure to industrial materials, processes, and stresses on the potential for illness, disease, or loss of functional capacity in aging adults.

Inspections.

(b) The Secretary of Health, Education, and Welfare is authorized to make inspections and question employers and employees as provided in section 8 of this Act in order to carry out his functions and responsibilities under this section.

Contract authority.

(c) The Secretary is authorized to enter into contracts, agreements, or other arrangements with appropriate public agencies or private organizations for the purpose of conducting studies relating to his responsibilities under this Act. In carrying out his responsibilities

84 STAT. 1612

under this subsection, the Secretary shall cooperate with the Secretary of Health, Education, and Welfare in order to avoid any duplication of efforts under this section.

(d) Information obtained by the Secretary and the Secretary of Health, Education, and Welfare under this section shall be disseminated by the Secretary to employers and employees and organizations thereof.

(e) The functions of the Secretary of Health, Education, and Welfare under this Act shall, to the extent feasible, be delegated to the Director of the National Institute for Occupational Safety and Health established by section 22 of this Act.

Delegation of functions.

TRAINING AND EMPLOYEE EDUCATION

SEC. 21. (a) The Secretary of Health, Education, and Welfare, after consultation with the Secretary and with other appropriate Federal departments and agencies, shall conduct, directly or by grants or contracts (1) education programs to provide an adequate supply of qualified personnel to carry out the purposes of this Act, and (2) informational programs on the importance of and proper use of adequate safety and health equipment.

(b) The Secretary is also authorized to conduct, directly or by grants or contracts, short-term training of personnel engaged in work related to his responsibilities under this Act.

(c) The Secretary, in consultation with the Secretary of Health, Education, and Welfare, shall (1) provide for the establishment and supervision of programs for the education and training of employers and employees in the recognition, avoidance, and prevention of unsafe or unhealthful working conditions in employments covered by this Act, and (2) consult with and advise employers and employees, and organizations representing employers and employees as to effective means of preventing occupational injuries and illnesses.

NATIONAL INSTITUTE FOR OCCUPATIONAL SAFETY AND HEALTH

SEC. 22. (a) It is the purpose of this section to establish a National Institute for Occupational Safety and Health in the Department of Health, Education, and Welfare in order to carry out the policy set forth in section 2 of this Act and to perform the functions of the Secretary of Health, Education, and Welfare under sections 20 and 21 of this Act.

Establishment.

(b) There is hereby established in the Department of Health, Education, and Welfare a National Institute for Occupational Safety and Health. The Institute shall be headed by a Director who shall be appointed by the Secretary of Health, Education, and Welfare, and who shall serve for a term of six years unless previously removed by the Secretary of Health, Education, and Welfare.

Director, appointment, term.

(c) The Institute is authorized to—

(1) develop and establish recommended occupational safety and health standards; and

(2) perform all functions of the Secretary of Health, Education, and Welfare under sections 20 and 21 of this Act.

(d) Upon his own initiative, or upon the request of the Secretary or the Secretary of Health, Education, and Welfare, the Director is authorized (1) to conduct such research and experimental programs as he determines are necessary for the development of criteria for new and improved occupational safety and health standards, and (2) after

84 STAT. 1613

consideration of the results of such research and experimental programs make recommendations concerning new or improved occupational safety and health standards. Any occupational safety and health standard recommended pursuant to this section shall immediately be forwarded to the Secretary of Labor, and to the Secretary of Health, Education, and Welfare.

(e) In addition to any authority vested in the Institute by other provisions of this section, the Director, in carrying out the functions of the Institute. is authorized to—

(1) prescribe such regulations as he deems necessary governing the manner in which its functions shall be carried out;

(2) receive money and other property donated, bequeathed, or devised. without condition or restriction other than that it be used for the purposes of the Institute and to use, sell, or otherwise dispose of such property for the purpose of carrying out its functions;

(3) receive (and use, sell, or otherwise dispose of, in accordance with paragraph (2)), money and other property donated, bequeathed, or devised to the Institute with a condition or restriction. including a condition that the Institute use other funds of the Institute for the purposes of the gift;

(4) in accordance with the civil service laws, appoint and fix the compensation of such personnel as may be necessary to carry out the provisions of this section;

(5) obtain the services of experts and consultants in accordance with the provisions of section 3109 of title 5, United States Code;

80 Stat. 416.

(6) accept and utilize the services of voluntary and noncompensated personnel and reimburse them for travel expenses, including per diem, as authorized by section 5703 of title 5, United States Code;

83 Stat. 190.

(7) enter into contracts, grants or other arrangements, or modifications thereof to carry out the provisions of this section, and such contracts or modifications thereof may be entered into without performance or other bonds, and without regard to section 3709 of the Revised Statutes, as amended (41 U.S.C. 5), or any other provision of law relating to competitive bidding;

(8) make advance, progress, and other payments which the Director deems necessary under this title without regard to the provisions of section 3648 of the Revised Statutes, as amended (31 U.S.C. 529) ; and

(9) make other necessary expenditures.

Annual report
to HEW,
President, and
Congress.

(f) The Director shall submit to the Secretary of Health, Education, and Welfare, to the President, and to the Congress an annual report of the operations of the Institute under this Act, which shall include a detailed statement of all private and public funds received and expended by it, and such recommendations as he deems appropriate.

GRANTS TO THE STATES

SEC. 23. (a) The Secretary is authorized, during the fiscal year ending June 30, 1971, and the two succeeding fiscal years, to make grants to the States which have designated a State agency under section 18 to assist them—

(1) in identifying their needs and responsibilities in the area of occupational safety and health,

(2) in developing State plans under section 18, or

December 29, 1970   - 25 -   Pub. Law 91-596

84 STAT. 1614

(3) in developing plans for—

(A) establishing systems for the collection of information concerning the nature and frequency of occupational injuries and diseases;

(B) increasing the expertise and enforcement capabilities of their personnel engaged in occupational safety and health programs; or

(C) otherwise improving the administration and enforcement of State occupational safety and health laws, including standards thereunder, consistent with the objectives of this Act.

(b) The Secretary is authorized, during the fiscal year ending June 30, 1971, and the two succeeding fiscal years, to make grants to the States for experimental and demonstration projects consistent with the objectives set forth in subsection (a) of this section.

(c) The Governor of the State shall designate the appropriate State agency for receipt of any grant made by the Secretary under this section.

(d) Any State agency designated by the Governor of the State desiring a grant under this section shall submit an application therefor to the Secretary.

(e) The Secretary shall review the application, and shall, after consultation with the Secretary of Health, Education, and Welfare, approve or reject such application.

(f) The Federal share for each State grant under subsection (a) or (b) of this section may not exceed 90 per centum of the total cost of the application. In the event the Federal share for all States under either such subsection is not the same, the differences among the States shall be established on the basis of objective criteria.

(g) The Secretary is authorized to make grants to the States to assist them in administering and enforcing programs for occupational safety and health contained in State plans approved by the Secretary pursuant to section 18 of this Act. The Federal share for each State grant under this subsection may not exceed 50 per centum of the total cost to the State of such a program. The last sentence of subsection (f) shall be applicable in determining the Federal share under this subsection.

(h) Prior to June 30, 1973, the Secretary shall, after consultation with the Secretary of Health, Education, and Welfare, transmit a report to the President and to the Congress, describing the experience under the grant programs authorized by this section and making any recommendations he may deem appropriate.   Report to President and Congress.

STATISTICS

SEC. 24. (a) In order to further the purposes of this Act, the Secretary, in consultation with the Secretary of Health, Education, and Welfare, shall develop and maintain an effective program of collection, compilation, and analysis of occupational safety and health statistics. Such program may cover all employments whether or not subject to any other provisions of this Act but shall not cover employments excluded by section 4 of the Act. The Secretary shall compile accurate statistics on work injuries and illnesses which shall include all disabling, serious, or significant injuries and illnesses, whether or not involving loss of time from work, other than minor injuries requiring only first aid treatment and which do not involve medical treatment, loss of consciousness, restriction of work or motion, or transfer to another job.

(b) To carry out his duties under subsection (a) of this section, the Secretary may—

(1) promote, encourage, or directly engage in programs of studies, information and communication concerning occupational safety and health statistics;

(2) make grants to States or political subdivisions thereof in order to assist them in developing and administering programs dealing with occupational safety and health statistics; and

(3) arrange, through grants or contracts, for the conduct of such research and investigations as give promise of furthering the objectives of this section.

(c) The Federal share for each grant under subsection (b) of this section may be up to 50 per centum of the State's total cost.

(d) The Secretary may, with the consent of any State or political subdivision thereof, accept and use the services, facilities, and employees of the agencies of such State or political subdivision, with or without reimbursement, in order to assist him in carrying out his functions under this section.

Reports.

(e) On the basis of the records made and kept pursuant to section 8(c) of this Act, employers shall file such reports with the Secretary as he shall prescribe by regulation, as necessary to carry out his functions under this Act.

(f) Agreements between the Department of Labor and States pertaining to the collection of occupational safety and health statistics already in effect on the effective date of this Act shall remain in effect until superseded by grants or contracts made under this Act.

AUDITS

SEC. 25. (a) Each recipient of a grant under this Act shall keep such records as the Secretary or the Secretary of Health, Education, and Welfare shall prescribe, including records which fully disclose the amount and disposition by such recipient of the proceeds of such grant, the total cost of the project or undertaking in connection with which such grant is made or used, and the amount of that portion of the cost of the project or undertaking supplied by other sources, and such other records as will facilitate an effective audit.

(b) The Secretary or the Secretary of Health, Education, and Welfare, and the Comptroller General of the United States, or any of their duly authorized representatives, shall have access for the purpose of audit and examination to any books, documents, papers, and records of the recipients of any grant under this Act that are pertinent to any such grant.

ANNUAL REPORT

SEC. 26. Within one hundred and twenty days following the convening of each regular session of each Congress, the Secretary and the Secretary of Health, Education, and Welfare shall each prepare and submit to the President for transmittal to the Congress a report upon the subject matter of this Act, the progress toward achievement of the purpose of this Act, the needs and requirements in the field of occupational safety and health, and any other relevant information. Such reports shall include information regarding occupational safety and health standards, and criteria for such standards, developed during the preceding year; evaluation of standards and criteria previously developed under this Act, defining areas of emphasis for new criteria and standards; an evaluation of the degree of observance of applicable occupational safety and health standards, and a summary

December 29, 1970   - 27 -   Pub. Law 91-596

of inspection and enforcement activity undertaken; analysis and evaluation of research activities for which results have been obtained under governmental and nongovernmental sponsorship; an analysis of major occupational diseases; evaluation of available control and measurement technology for hazards for which standards or criteria have been developed during the preceding year; description of cooperative efforts undertaken between Government agencies and other interested parties in the implementation of this Act during the preceding year; a progress report on the development of an adequate supply of trained manpower in the field of occupational safety and health, including estimates of future needs and the efforts being made by Government and others to meet those needs; listing of all toxic substances in industrial usage for which labeling requirements, criteria, or standards have not yet been established; and such recommendations for additional legislation as are deemed necessary to protect the safety and health of the worker and improve the administration of this Act.

NATIONAL COMMISSION ON STATE WORKMEN'S COMPENSATION LAWS

SEC. 27. (a) (1) The Congress hereby finds and declares that—

(A) the vast majority of American workers, and their families, are dependent on workmen's compensation for their basic economic security in the event such workers suffer disabling injury or death in the course of their employment; and that the full protection of American workers from job-related injury or death requires an adequate, prompt, and equitable system of workmen's compensation as well as an effective program of occupational health and safety regulation; and

(B) in recent years serious questions have been raised concerning the fairness and adequacy of present workmen's compensation laws in the light of the growth of the economy, the changing nature of the labor force, increases in medical knowledge, changes in the hazards associated with various types of employment, new technology creating new risks to health and safety, and increases in the general level of wages and the cost of living.

(2) The purpose of this section is to authorize an effective study and objective evaluation of State workmen's compensation laws in order to determine if such laws provide an adequate, prompt, and equitable system of compensation for injury or death arising out of or in the course of employment.

(b) There is hereby established a National Commission on State Workmen's Compensation Laws.   Establishment.

(c) (1) The Workmen's Compensation Commission shall be com-   Membership.
posed of fifteen members to be appointed by the President from among members of State workmen's compensation boards, representatives of insurance carriers, business, labor, members of the medical profession having experience in industrial medicine or in workmen's compensation cases, educators having special expertise in the field of workmen's compensation, and representatives of the general public. The Secretary, the Secretary of Commerce, and the Secretary of Health, Education, and Welfare shall be ex officio members of the Workmen's Compensation Commission:

(2) Any vacancy in the Workmen's Compensation Commission shall not affect its powers.

(3) The President shall designate one of the members to serve as Chairman and one to serve as Vice Chairman of the Workmen's Compensation Commission.

Pub. Law 91-596   - 28 -   December 29, 1970

84 STAT. 1617

Quorum.

(4) Eight members of the Workmen's Compensation Commission shall constitute a quorum.

Study.

(d)(1) The Workmen's Compensation Commission shall undertake a comprehensive study and evaluation of State workmen's compensation laws in order to determine if such laws provide an adequate, prompt, and equitable system of compensation. Such study and evaluation shall include, without being limited to, the following subjects: (A) the amount and duration of permanent and temporary disability benefits and the criteria for determining the maximum limitations thereon, (B) the amount and duration of medical benefits and provisions insuring adequate medical care and free choice of physician, (C) the extent of coverage of workers, including exemptions based on numbers or type of employment, (D) standards for determining which injuries or diseases should be deemed compensable, (E) rehabilitation, (F) coverage under second or subsequent injury funds, (G) time limits on filing claims, (H) waiting periods, (I) compulsory or elective coverage, (J) administration, (K) legal expenses, (L) the feasibility and desirability of a uniform system of reporting information concerning job-related injuries and diseases and the operation of workmen's compensation laws, (M) the resolution of conflict of laws, extraterritoriality and similar problems arising from claims with multistate aspects, (N) the extent to which private insurance carriers are excluded from supplying workmen's compensation coverage and the desirability of such exclusionary practices, to the extent they are found to exist, (O) the relationship between workmen's compensation on the one hand, and old-age, disability, and survivors insurance and other types of insurance, public or private, on the other hand, (P) methods of implementing the recommendations of the Commission.

Report to President and Congress.

(2) The Workmen's Compensation Commission shall transmit to the President and to the Congress not later than July 31, 1972, a final report containing a detailed statement of the findings and conclusions of the Commission, together with such recommendations as it deems advisable.

Hearings.

(e)(1) The Workmen's Compensation Commission or, on the authorization of the Workmen's Compensation Commission, any subcommittee or members thereof, may, for the purpose of carrying out the provisions of this title, hold such hearings, take such testimony, and sit and act at such times and places as the Workmen's Compensation Commission deems advisable. Any member authorized by the Workmen's Compensation Commission may administer oaths or affirmations to witnesses appearing before the Workmen's Compensation Commission or any subcommittee or members thereof.

(2) Each department, agency, and instrumentality of the executive branch of the Government, including independent agencies, is authorized and directed to furnish to the Workmen's Compensation Commission, upon request made by the Chairman or Vice Chairman, such information as the Workmen's Compensation Commission deems necessary to carry out its functions under this section.

(f) Subject to such rules and regulations as may be adopted by the Workmen's Compensation Commission, the Chairman shall have the power to—

(1) appoint and fix the compensation of an executive director, and such additional staff personnel as he deems necessary, without regard to the provisions of title 5, United States Code, governing appointments in the competitive service, and without regard to the provisions of chapter 51 and subchapter III of chapter 53 of such title relating to classification and General Schedule

80 Stat. 378.
5 USC 101.

5 USC 5101,
5331.

December 29, 1970   - 29 -   **Pub. Law 91-596**

84 STAT. 1618

pay rates, but at rates not in excess of the maximum rate for GS-18 of the General Schedule under section 5332 of such title, and
Ante, p. 198-1.

(2) procure temporary and intermittent services to the same extent as is authorized by section 3109 of title 5, United States Code.

(g) The Workmen's Compensation Commission is authorized to enter into contracts with Federal or State agencies, private firms, institutions, and individuals for the conduct of research or surveys, the preparation of reports, and other activities necessary to the discharge of its duties.
80 Stat. 416.
Contract
authorization.

(h) Members of the Workmen's Compensation Commission shall receive compensation for each day they are engaged in the performance of their duties as members of the Workmen's Compensation Commission at the daily rate prescribed for GS-18 under section 5332 of title 5, United States Code, and shall be entitled to reimbursement for travel, subsistence, and other necessary expenses incurred by them in the performance of their duties as members of the Workmen's Compensation Commission.
Compensation;
travel ex-
penses.

(i) There are hereby authorized to be appropriated such sums as may be necessary to carry out the provisions of this section.
Appropriation.

(j) On the ninetieth day after the date of submission of its final report to the President, the Workmen's Compensation Commission shall cease to exist.
Termination.

ECONOMIC ASSISTANCE TO SMALL BUSINESSES

SEC. 28. (a) Section 7(b) of the Small Business Act, as amended, is amended—
72 Stat. 387;
83 Stat. 802.
15 USC 636.

(1) by striking out the period at the end of "paragraph (5)" and inserting in lieu thereof "; and"; and

(2) by adding after paragraph (5) a new paragraph as follows:

"(6) to make such loans (either directly or in cooperation with banks or other lending institutions through agreements to participate on an immediate or deferred basis) as the Administration may determine to be necessary or appropriate to assist any small business concern in effecting additions to or alterations in the equipment, facilities, or methods of operation of such business in order to comply with the applicable standards promulgated pursuant to section 6 of the Occupational Safety and Health Act of 1970 or standards adopted by a State pursuant to a plan approved under section 18 of the Occupational Safety and Health Act of 1970, if the Administration determines that such concern is likely to suffer substantial economic injury without assistance under this paragraph."

(b) The third sentence of section 7(b) of he Small Business Act, as amended, is amended by striking out "or (5)" after "paragraph (3)" and inserting a comma followed by "(5) or (6)".

(c) Section 4(c)(1) of the Small Business Act, as amended, is amended by inserting "7(b)(6)," after "7(b)(5),".
80 Stat. 132.
15 USC 633.

(d) Loans may also be made or guaranteed for the purposes set forth in section 7(b)(6) of the Small Business Act, as amended, pursuant to the provisions of section 202 of the Public Works and Economic Development Act of 1965, as amended.
79 Stat. 556.
42 USC 3142.

ADDITIONAL ASSISTANT SECRETARY OF LABOR

SEC. 29. (a) Section 2 of the Act of April 17, 1946 (60 Stat. 91) as amended (29 U.S.C. 553) is amended by—
75 Stat. 338.

(1) striking out "four" in the first sentence of such section and inserting in lieu thereof "five"; and

(2) adding at the end thereof the following new sentence, "One of such Assistant Secretaries shall be an Assistant Secretary of Labor for Occupational Safety and Health.".

80 Stat. 462.   (b) Paragraph (20) of section 5315 of title 5, United States Code, is amended by striking out "(4)" and inserting in lieu thereof "(5)".

ADDITIONAL POSITIONS

SEC. 30. Section 5108(c) of title 5, United States Code, is amended by—

(1) striking out the word "and" at the end of paragraph (8);

(2) striking out the period at the end of paragraph (9) and inserting in lieu thereof a semicolon and the word "and"; and

(3) by adding immediately after paragraph (9) the following new paragraph:

"(10) (A) the Secretary of Labor, subject to the standards and procedures prescribed by this chapter, may place an additional twenty-five positions in the Department of Labor in GS–16, 17, and 18 for the purposes of carrying out his responsibilities under the Occupational Safety and Health Act of 1970;

"(B) the Occupational Safety and Health Review Commission, subject to the standards and procedures prescribed by this chapter, may place ten positions in GS–16, 17, and 18 in carrying out its functions under the Occupational Safety and Health Act of 1970."

EMERGENCY LOCATOR BEACONS

72 Stat. 775.   SEC. 31. Section 601 of the Federal Aviation Act of 1958 is amended
49 USC 1421.   by inserting at the end thereof a new subsection as follows:

"EMERGENCY LOCATOR BEACONS

"(d) (1) Except with respect to aircraft described in paragraph (2) of this subsection, minimum standards pursuant to this section shall include a requirement that emergency locator beacons shall be installed—

"(A) on any fixed-wing, powered aircraft for use in air commerce the manufacture of which is completed, or which is imported into the United States, after one year following the date of enactment of this subsection; and

"(B) on any fixed-wing, powered aircraft used in air commerce after three years following such date.

"(2) The provisions of this subsection shall not apply to jet-powered aircraft; aircraft used in air transportation (other than air taxis and charter aircraft); military aircraft; aircraft used solely for training purposes not involving flights more than twenty miles from its base; and aircraft used for the aerial application of chemicals."

SEPARABILITY

SEC. 32. If any provision of this Act, or the application of such provision to any person or circumstance, shall be held invalid, the remainder of this Act, or the application of such provision to persons or circumstances other than those as to which it is held invalid, shall not be affected thereby.

December 29, 1970   - 31 -   Pub. Law 91-596

84 STAT. 1620

APPROPRIATIONS

Sec. 33. There are authorized to be appropriated to carry out this Act for each fiscal year such sums as the Congress shall deem necessary.

EFFECTIVE DATE

Sec. 34. This Act shall take effect one hundred and twenty days after the date of its enactment.

Approved December 29, 1970.

○

LEGISLATIVE HISTORY:

HOUSE REPORTS: No. 91-1291 accompanying H.R. 16785 (Comm. on
               Education and Labor) and No. 91-1765 (Comm. of
               Conference).
SENATE REPORT No. 91-1282 (Comm. on Labor and Public Welfare).
CONGRESSIONAL RECORD, Vol. 116 (1970):
   Oct. 13, Nov. 16, 17, considered and passed Senate.
   Nov. 23, 24, considered and passed House, amended, in lieu
             of H.R. 16785.
   Dec. 16, Senate agreed to conference report.
   Dec. 17, House agreed to conference report.

U.S. GOVERNMENT PRINTING OFFICE : 1986 O --155-936

OSHA REGIONAL OFFICES

Correspondence to OSHA Regional Offices should be addressed to the U.S. Department of Labor, Occupational Safety and Health Administration, at the following addresses:

REGION	CITY	ADDRESS	TELEPHONE
I	Boston	16–18 North Street 1 Dock Square Building 4th Floor Boston, MA 02109	(617) 223–6710
II	New York	Room 3445, 1 Astor Plaza 1515 Broadway New York, NY 10036	(212) 944–3432
III	Philadelphia	Gateway Building Suite 2100 3535 Market Street Philadelphia, PA 19104	(215) 596–1201
IV	Atlanta	1375 Peachtree St., N.E. Suite 587 Atlanta, GA 30367	(404) 881–3573
V	Chicago	230 South Dearborn Street 32nd Floor, Room 3244 Chicago, IL 60604	(312) 353–2220
VI	Dallas	525 Griffin Square Room 602 Dallas, TX 75202	(214) 767–4731
VII	Kansas City	911 Walnut Street Room 406 Kansas City, MO 64106	(816) 374–5861
VIII	Denver	Federal Building Room 1554 1961 Stout Street Denver, CO 80294	(303) 844–3061
IX	San Francisco	9470 Federal Building 450 Golden Gate Avenue P.O. Box 36017 San Francisco, CA 94102	(415) 556–7260
X	Seattle	Federal Office Building Room 6003 909 First Avenue Seattle, WA 98174	(206) 442–5930

USEFUL OSHA BOOKLET PUBLICATION

OSHA NUMBER	TITLE
2056	All About OSHA
2098	OSHA Inspections
2226	Excavating and Trenching Operations
2227	Essentials of Machine Guarding
2231	Organizing a Safety Committee
2236	Material Handling and Storage
2237	Handling Hazardous Materials
2253	Workers' Rights Under OSHA
2288	Investigating Accidents in the Workplace
3007	Ground-Fault Protection on Construction Sites
3047	Consultation Service For the Employer
3071	Job Hazard Analysis
3075	Controlling Electrical Hazards
3077	Personal Protective Equipment
3079	Respiratory Protection
3080	Hand and Power Tools
3088	How to Prepare for Workplace Emergencies
3092	Working Safely With Video Display Terminals
3093	Workplace Health Programs
3097	Electrical Standards for Construction

Note: For other OSHA publications, contact the OSHA office in your area or State.

FALL PREVENTION REQUIREMENTS

SAFETY BELTS (OSHA 1926.104)
Working above 6 feet.
SAFETY NETS (OSHA 1926.105)
Working above 25 feet.
LIFE JACKETS (OSHA 1926.106)
Working over or near water.
SCAFFOLDS (OSHA 1926.451)
Guardrails and toeboards required when scaffolds are installed more than 10 feet above ground.
CATCH PLATFORM (OSHA 1926.451)
A catch platform shall be installed below the working area of roofs more than 16 feet from ground to eaves with a roof slope greater than 4 inches in 12 inches.
FLOOR OPENINGS (OSHA 1926.500)
Floor openings shall be guarded by a standard railing or cover.
WALL OPENINGS (OSHA 1926.500)
Wall openings shall be guarded by railings when a drop is more than 4 feet.
OPEN SIDED FLOORS, PLATFORMS AND RUNWAYS (OSHA 1926.500)
Open sides shall be guarded by railings when a drop is more than 6 feet.
STAIRWAY (OSHA 1926.500)
All stairways with four or more risers shall be equipped with a railing.
LOW PITCH ROOF (OSHA 1926.500)
Fall protection is required when height from roof to ground exceeds 16 feet by one of the following:

• Motion stopping safety system
• Warning lines 6 feet from edge of roof
• Safety monitor (competent person)
• Safety belt system

TEMPORARY FLOORING (OSHA 1926.700)
One-half inch wire rope shall be installed on all sides of temporary flooring. The worker shall be protected by a safety belt, lifeline and anchorage.

Note: For more specifics, consult the above OSHA requirements.

CALCULATING INJURY FREQUENCY AND SEVERITY RATES

Calculating injury frequency rate shows the number of injuries that your company has suffered for every 200,000 man hours worked.

EXAMPLE: Your employees have worked 2,000,000 man hours and have suffered 20 lost-time injuries

$$\frac{20 \,(\text{injuries} \times 200,000 \,(\text{man hours})}{2,000,000 \,(\text{man hours worked})} = 2 \text{ injuries per 100 workers}$$

NOTE: The number 2 represents the number of injuries for every 200,000 man hours worked.

Calculating the severity rate shows the number of days that your company has lost for every 200,000 man hours worked.

EXAMPLE: Your employees have worked 2,000,000 man hours and have lost 120 man days

$$\frac{120 \,(\text{lost man days}) \times 200,000 \,(\text{man hours})}{2,000,000 \,(\text{man hours worked})} = 12 \text{ lost man days per 100 workers}$$

NOTE: The number 12 represents the number of days lost for every 200,000 man hours worked.

The 200,000 constant in the above formula represents 100 workers that have worked a combined 200,000 hours based on a 40 hour work week for 50 weeks:

$$100 \times 40 \times 50 = 200,000 \text{ man hours}$$

LADDER SAFETY

- **EXTENSION LADDERS** — Do not extend the ladder as far as it will go; leave at least 3 feet. That will allow a margin of safety in the middle, where the two sections overlap.
- **PLACEMENT OF LADDER** — The base of the ladder should be placed away from a building about a quarter the distance of its length. At a height of 20 feet, for instance, the base of the ladder would be 5 feet from the house. The ladder should extend 3 feet above a roof, when using the ladder to climb on to a roof.
- **STEP LADDER** — Do not stand on the top flat of a step ladder. That should only be used as a work surface.
- **SAFETY HINTS** — Maintain good balance by not over-reaching to the sides. Use both hands when climbing the ladder.
- **RAISING THE LADDER** — To get the ladder in an upright position, push the foot of the ladder against the building. Now go backwards to the top part of the ladder — grab the top rung and lift it to your chin; then walk under the ladder, moving your grip to other rungs as you slowly bring the ladder in a vertical position. Last, position the feet of the ladder as stated above.

OSHA'S 25 MOST COMMON WORKPLACE SAFETY VIOLATIONS

- Posting the OSHA Notice of Employee Obligation.
- Grounding electrical equipment connected by cord and plug.
- Machinery guards at point of operation.
- Guards for pulleys.
- Guards for live part of electrical equipment.
- Enclosures for blades of fans in use less than 7 feet above floor.
- Guards for belt, rope, and chain drives.
- Housekeeping: orderliness in aisles and passageways.
- Guards for vertical and inclined belts.
- Maintaining a log and summary of job injuries and illnesses.
- Design of guards for abrasive wheel machinery.
- Guard adjustments for abrasive wheel machinery.
- Enclosures for sprocket wheel and chains.
- Guardrails for open-sided floors, machinery, platforms, runways, 4 feet or more above adjacent floor or ground level.
- Personal protective equipment for eyes and face.
- Construction sites: equipment grounding conductors or GFCI protection.
- Flexible electrical cords and cables.
- Securely clamping workrests for abrasive wheel machinery.
- Construction sites: guardrails for open-sided floors (etc.).
- Work area facilities, where appropriate, for quick drenching or flushing of eyes and body.
- Repairs for powered industrial trucks.
- Using compressed air for cleaning purposes.
- Anchoring fixed machinery.
- Protection against effects of noise exposure.
- Marking and storing cylinders in welding, cutting, and brazing.

LIST OF POSSIBLE PROBLEMS TO BE INSPECTED

Acids	Elevators	Ramps
Aisles	Explosives	Raw materials
Alarms	Extinguishers	Respirators
Atmospheres	Eye protection	Roads
Automobiles	Flammables	Roofs
Barrels	Floors	Ropes
Bins	Fork lifts	Safety devices
Blinker lights	Fumes	Safety shoes
Boilers	Gas cylinders	Scaffolds
Borers	Gas engines	Shafts
Buggies	Gases	Shapers
Buildings	Hand tools	Shelves
Cabinets	Hard hats	Sirens
Cables	Hoist	Slings
Carboys	Horns and signals	Solvents and lockers
Catwalks	Hoses	Spray
Caustics	Hydrants	Sprinkler systems
Chemicals	Ladders	Stairs
Closets	Lathes	Steam engines
Connectors	Lights	Sumps
Containers	Mills	Switches
Controls	Mists	Tanks
Conveyors	Motorized carts	Trucks
Cranes	Piping	Vats
Crossing lights	Pits	Walkways
Cutters	Platforms	Walls
Docks	Power tools	Warning devices
Doors	Presses	
Dust	Racks	
Electric motors	Railroad cars	

LIST OF FREQUENTLY CITED OSHA STANDARDS (GENERAL)

TITLE	OSHA PART NUMBER
Safety/health policy	1903.1
OSHA poster	1903.2
Employee complaint	1903.11
Recordkeeping	1904.2
Material Safety Data Sheet	1910.20
Walk/work Surfaces	1910.21
Housekeeping	1910.22
Guarded openings	1910.23
Stairs	1910.24
Portable wood ladders	1910.25
Portable metal ladders	1910.26
Fixed ladders	1910.27
Scaffolding	1910.28
Loading docks	1910.30
Means of egress	1910.35
Emergency plans	1910.38
Powered platforms	1910.66
Ventilation	1910.94
Noise exposure	1910.95
LASERs/microwaves	1910.97
Hazardous materials	1910.101
Flammables	1910.106
Dip tank solvents	1910.108
Explosives	1910.109
LPG gases	1910.110
Ammonia	1910.111
Personal protective equipment	1910.132
Respiratory protection	1910.134
Sanitation	1910.141
Safety color markings	1910.144
Signs/posters/tags	1910.145
Eyewash/showers/medical	1910.151
Fire prevention/protection	1910.155
Compressed gases	1910.166
Weight handling equipment	1910.176
Cranes	1910.179
Material handling	1910.189
Machine guards	1910.211
Grinding wheels	1910.215
Hand and power tools	1910.241
Welding/cutting	1910.251
Confined spaces	1910.252
Special industries	1910.261
Electrical safety	1910.301
Toxic/hazardous substances	1910.1000
Hazard communication	1910.1200

LIST OF FREQUENTLY CITED OSHA STANDARDS (CONSTRUCTION)

TITLE	OSHA PART NUMBER
Safety/health policy	1903.1
OSHA poster	1903.2
Employee complaint	1903.11
Record-keeping	1904.2
Rules of construction	1926.16
Contractor requirements and enforcement	1926.20
Safety training	1926.21
Recording and reporting injuries	1926.22
First aid and medical attention	1926.23 & .50
Fire protection and prevention	1926.24 & .150
Housekeeping	1926.25
Illumination	1926.26 & .56
Sanitation	1926.27 & .51
Personal protective equipment	1926.28 & .100
Noise exposure	1926.52
LASERs	1926.54
Gases, vapors, fumes, dust, mists	1926.55
Ventilation	1926.57
Respiratory protection	1926.103
Safety belts, lifelines and lanyards	1926.104
Safety nets	1926.105
Signs, signals and barricades	1926.200
Material handling, storage, etc.	1926.250
Hand and power tools	1926.300
Welding and cutting	1926.350
Electrical	1926.400
Ladders and scaffolding	1926.450
Floor and wall openings and stairways	1926.500
Cranes, hoists, elevators, conveyors, etc.	1926.550
Motor vehicles, etc.	1926.600
Excavations, trenching and shoring	1926.650
Concrete, forms, and shoring	1926.700
Steel erection	1926.750
Tunnels, shafts, caissons, etc.	1926.800
Demolition	1926.850
Blasting and explosives	1926.900
Power transmission and distribution	1926.950
Rollover and overhead protection	1926.1000

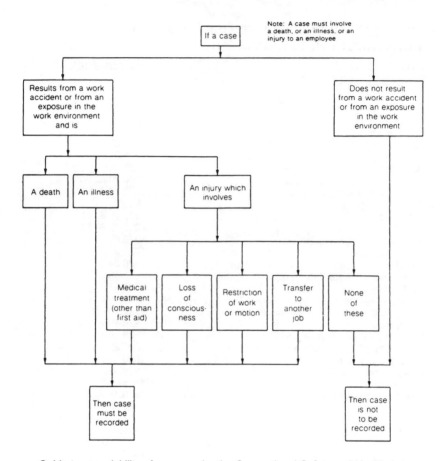

Guide to recordability of cases under the Occupational Safety and Health Act.

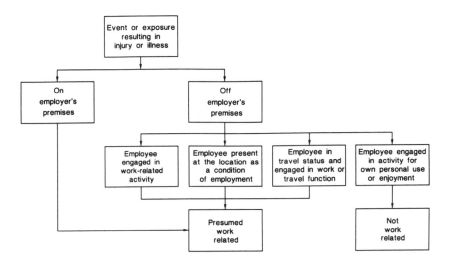

Guidelines for establishing work relationships.

Bureau of Labor Statistics
Log and Summary of Occupational
Injuries and Illnesses

NOTE:	This form is required by Public Law 91-596 and must be kept in the establishment for 5 years. Failure to maintain and post can result in the issuance of citations and assessment of penalties. *(See posting requirements on the other side of form.)*	**RECORDABLE CASES:** You are required to record information about every occupational **death**; every nonfatal occupational **illness**; and those nonfatal occupational **injuries** which involve one or more of the following: loss of consciousness, restriction of work or motion, transfer to another job, or medical treatment (other than first aid). *(See definitions on the other side of form.)*

Case or File Number	Date of Injury or Onset of Illness	Employee's Name	Occupation	Department	Description of Injury or Illness
Enter a nonduplicating number which will facilitate comparisons with supplementary records.	Enter Mo./day.	Enter first name or initial, middle initial, last name.	Enter regular job title, not activity employee was performing when injured or at onset of illness. In the absence of a formal title, enter a brief description of the employee's duties.	Enter department in which the employee is regularly employed or a description of normal workplace to which employee is assigned, even though temporarily working in another department at the time of injury or illness.	Enter a brief description of the injury or illness and indicate the part or parts of body affected. Typical entries for this column might be Amputation of 1st joint right forefinger; Strain of lower back; Contact dermatitis on both hands; Electrocution—body.
(A)	(B)	(C)	(D)	(E)	(F)
					PREVIOUS PAGE TOTALS ➡
					TOTALS (instructions on other side of form.) ➡

OSHA No. 200

U.S. Department of Labor

For Calendar Year 19 ______ Page ___ of ___

Company Name

Establishment Name

Establishment Address

Form Approved
O.M.B. No. 1220-0029

Extent of and Outcome of INJURY						Type, Extent of, and Outcome of ILLNESS												
Fatalities	Nonfatal Injuries					Type of Illness							Fatalities	Nonfatal Illnesses				
Injury Related	Injuries With Lost Workdays				Injuries Without Lost Workdays	CHECK Only One Column for Each Illness *(See other side of form for terminations or permanent transfers.)*							Illness Related	Illnesses With Lost Workdays				Illnesses Without Lost Workdays
Enter **DATE** of death. Mo./day/yr.	Enter a **CHECK** if injury involves days away from work, or days of restricted work activity, or both.	Enter a **CHECK** if injury involves days away from work.	Enter number of **DAYS** *away from work.*	Enter number of **DAYS** of *restricted work activity.*	Enter a **CHECK** if no entry was made in columns 1 or 2 but the injury is recordable as defined above.	Occupational skin diseases or disorders	Dust diseases of the lungs	Respiratory conditions due to toxic agents	Poisoning (systemic effects of toxic materials)	Disorders due to physical agents	Disorders associated with repeated trauma	All other occupational illnesses	Enter **DATE** of death. Mo./day/yr.	Enter a **CHECK** if illness involves days away from work, or days of restricted work activity, or both.	Enter a **CHECK** if illness involves days away from work.	Enter number of **DAYS** *away from work.*	Enter number of **DAYS** of *restricted work activity.*	Enter a **CHECK** if no entry was made in columns 8 or 9.
(1)	(2)	(3)	(4)	(5)	(6)	(7)							(8)	(9)	(10)	(11)	(12)	(13)
						(a)	(b)	(c)	(d)	(e)	(f)	(g)						

INJURIES

ILLNESSES

Certification of Annual Summary Totals By ______________________ Title ______________________ Date ____________

FOLD

OSHA No. 200

POST ONLY THIS PORTION OF THE LAST PAGE NO LATER THAN FEBRUARY 1.

Instructions for OSHA No. 200

I. **Log and Summary of Occupational Injuries and Illnesses**

Each employer who is subject to the recordkeeping requirements of the Occupational Safety and Health Act of 1970 must maintain for each establishment a log of all recordable occupational injuries and illnesses. This form (OSHA No. 200) may be used for that purpose. A substitute for the OSHA No. 200 is acceptable if it is as detailed, easily readable, and understandable as the OSHA No. 200.

Enter each recordable case on the log within six (6) workdays after learning of its occurrence. Although other records must be maintained at the establishment to which they refer, it is possible to prepare and maintain the log at another location, using data processing equipment if desired. If the log is prepared elsewhere, a copy updated to within 45 calendar days must be present at all times in the establishment.

Logs must be maintained and retained for five (5) years following the end of the calendar year to which they relate. Logs must be available (normally at the establishment) for inspection and copying by representatives of the Department of Labor, or the Department of Health and Human Services, or States accorded jurisdiction under the Act. Access to the log is also provided to employees, former employees and their representatives.

II. **Changes in Extent of or Outcome of Injury or Illness**

If, during the 5-year period the log must be retained, there is a change in an extent and outcome of an injury or illness which affects entries in columns 1, 2, 6, 8, 9, or 13, the first entry should be lined out and a new entry made. For example, if an injured employee at first required only medical treatment but later lost workdays away from work, the check in column 6 should be lined out, and checks entered in columns 2 and 3 and the number of lost workdays entered in column 4.

In another example, if an employee with an occupational illness lost workdays, returned to work, and then died of the illness, any entries in columns 9 through 12 should be lined out and the date of death entered in column 8.

The entire entry for an injury or illness should be lined out if later found to be nonrecordable. For example: an injury which is later determined not to be work related, or which was initially thought to involve medical treatment but later was determined to have involved only first aid.

III. **Posting Requirements**

A copy of the totals and information following the fold line of the last page for the year must be posted at each establishment in the place or places where notices to employees are customarily posted. This copy must be posted no later than *February 1 and must remain in place until March 1.*

Even though there were no injuries or illnesses during the year, zeros must be entered on the totals line, and the form posted.

The person responsible for the *annual summary totals* shall certify that the totals are true and complete by signing at the bottom of the form.

IV. **Instructions for Completing Log and Summary of Occupational Injuries and Illnesses**

Column A — **CASE OR FILE NUMBER.** Self-explanatory.

Column B — DATE OF INJURY OR ONSET OF ILLNESS.
For occupational injuries, enter the date of the work accident which resulted in injury. For occupational illnesses, enter the date of initial diagnosis of illness, or, if absence from work occurred before diagnosis, enter the first day of the absence attributable to the illness which was later diagnosed or recognized.

Columns
C through F — Self-explanatory.

Columns
1 and 8 — INJURY OR ILLNESS-RELATED DEATHS.
Self-explanatory.

Columns
2 and 9 — INJURIES OR ILLNESSES WITH LOST WORKDAYS.
Self-explanatory.

Any injury which involves days away from work, or days of restricted work activity, or both must be recorded since it always involves one or more of the criteria for recordability.

Columns
3 and 10 — INJURIES OR ILLNESSES INVOLVING DAYS AWAY FROM WORK. Self-explanatory.

Columns
4 and 11 — LOST WORKDAYS—DAYS AWAY FROM WORK.
Enter the number of workdays (consecutive or not) on which the employee would have worked but could not because of occupational injury or illness. The number of lost workdays should not include the day of injury or onset of illness or any days on which the employee would not have worked even though able to work.
NOTE: For employees not having a regularly scheduled shift, such as certain truck drivers, construction workers, farm labor, casual labor, part-time employees, etc., it may be necessary to estimate the number of lost workdays. Estimates of lost workdays shall be based on prior work history of the employee AND days worked by employees, not ill or injured, working in the department and/or occupation of the ill or injured employee.

Columns
5 and 12 — LOST WORKDAYS—DAYS OF RESTRICTED WORK ACTIVITY.
Enter the number of workdays (consecutive or not) on which because of injury or illness:
(1) the employee was assigned to another job on a temporary basis, or
(2) the employee worked at a permanent job less than full time, or
(3) the employee worked at a permanently assigned job but could not perform all duties normally connected with it.

The number of lost workdays should not include the day of injury or onset of illness or any days on which the employee would not have worked even though able to work.

**Columns
6 and 13** — INJURIES OR ILLNESSES WITHOUT LOST
WORKDAYS. Self-explanatory.

**Columns 7a
through 7g** — TYPE OF ILLNESS.
Enter a check in only *one* column for each illness.

TERMINATION OR PERMANENT TRANSFER—Place an asterisk to
the right of the entry in columns 7a through 7g (type of illness) which
represented a termination of employment or permanent transfer.

V. Totals

Add number of entries in columns 1 and 8.
Add number of checks in columns 2, 3, 6, 7, 9, 10, and 13.
Add number of days in columns 4, 5, 11, and 12.
Yearly totals for each column (1-13) are required for posting. Running or
page totals may be generated at the discretion of the employer.

If an employee's loss of workdays is continuing at the time the totals are
summarized, estimate the number of future workdays the employee will
lose and add that estimate to the workdays already lost and include this
figure in the annual totals. No further entries are to be made with respect
to such cases in the next year's log.

VI. Definitions

OCCUPATIONAL INJURY is any injury such as a cut, fracture, sprain,
amputation, etc., which results from a work accident or from an éxpo-
sure involving a single incident in the work environment.
NOTE: Conditions resulting from animal bites, such as insect or snake
bites or from one-time exposure to chemicals, are considered to be injuries.

OCCUPATIONAL ILLNESS of an employee is any abnormal condition or
disorder, other than one resulting from an occupational injury, caused by
exposure to environmental factors associated with employment. It includes acute and chronic illnesses or diseases which may be caused by inhalation, absorption, ingestion, or direct contact.

The following listing gives the categories of occupational illnesses and disorders that will be utilized for the purpose of classifying recordable illnesses. For purposes of information, examples of each category are given.
These are typical examples, however, and are not to be considered the
complete listing of the types of illnesses and disorders that are to be count-
ed under each category.

7a.   Occupational Skin Diseases or Disorders
Examples: Contact dermatitis, eczema, or rash caused by primary irritants and sensitizers or poisonous plants; oil acne;
chrome ulcers; chemical burns or inflammations, etc.

7b.   Dust Diseases of the Lungs (Pneumoconioses)
Examples: Silicosis, asbestosis and other asbestos-related diseases, coal worker's pneumoconiosis, byssinosis, siderosis, and
other pneumoconioses.

7c.   Respiratory Conditions Due to Toxic Agents
Examples: Pneumonitis, pharyngitis, rhinitis or acute conges-
tion due to chemicals, dusts, gases, or fumes; farmer's lung, etc.

7d.   Poisoning (Systemic Effect of Toxic Materials)
Examples: Poisoning by lead, mercury, cadmium, arsenic, or
other metals; poisoning by carbon monoxide, hydrogen sulfide,
or other gases; poisoning by benzol, carbon tetrachloride, or
other organic solvents; poisoning by insecticide sprays such as
parathion, lead arsenate; poisoning by other chemicals such as
formaldehyde, plastics, and resins; etc.

7e.   Disorders Due to Physical Agents (Other than Toxic Materials)
Examples: Heatstroke, sunstroke, heat exhaustion, and other
effects of environmental heat; freezing, frostbite, and effects of
exposure to low temperatures; caisson disease; effects of ionizing
radiation (isotopes, X-rays, radium); effects of nonionizing radiation (welding flash, ultraviolet rays, microwaves, sunburn); etc.

7f.   Disorders Associated With Repeated Trauma
Examples: Noise-induced hearing loss; synovitis, tenosynovitis,
and bursitis; Raynaud's phenomena; and other conditions due to
repeated motion, vibration, or pressure.

7g.   All Other Occupational Illnesses
Examples: Anthrax, brucellosis, infectious hepatitis, malignant
and benign tumors, food poisoning, histoplasmosis, coccidioido-
mycosis, etc.

MEDICAL TREATMENT includes treatment (other than first aid) administered by a physician or by registered professional personnel under the
standing orders of a physician. Medical treatment does NOT include first-
aid treatment (one-time treatment and subsequent observation of minor
scratches, cuts, burns, splinters, and so forth, which do not ordinarily require medical care) even though provided by a physician or registered
professional personnel.

ESTABLISHMENT: A single physical location where business is conducted or where services or industrial operations are performed (for example:
a factory, mill, store, hotel, restaurant, movie theater, farm, ranch, bank,
sales office, warehouse, or central administrative office). Where distinctly
separate activities are performed at a single physical location such as construction activities operated from the same physical location as a lumber
yard, each activity shall be treated as a separate establishment.

For firms engaged in activities which may be physically dispersed, such as
agriculture; construction; transportation; communications; and electric,
gas, and sanitary services, records may be maintained at a place to which
employees report each day.

Records for personnel who do not primarily report or work at a single
establishment, such as traveling salesmen, technicians, engineers, etc., shall
be maintained at the location from which they are paid or the base from
which personnel operate to carry out their activities.

WORK ENVIRONMENT is comprised of the physical location, equipment,
materials processed or used, and the kinds of operations performed in the
course of an employee's work, whether on or off the employer's premises.

OSHA No. 101   Form approved
Case or File No. _____   OMB No. 44R 1453

Supplementary Record of Occupational Injuries and Illnesses

EMPLOYER

  1. Name _____

  2. Mail address _____
                (No. and street)   (City or town)   (State)

  3. Location, if different from mail address _____

INJURED OR ILL EMPLOYEE

  4. Name _____ Social Security No. _____
        (First name)   (Middle name)   (Last name)

  5. Home address _____
          (No. and street)   (City or town)   (State)

  6. Age _____   7. Sex: Male_____ Female_____ (Check one)

  8. Occupation _____
          (Enter regular job title, *not* the specific activity he was performing at time of injury.)

  9. Department _____
          (Enter name of department or division in which the injured person is regularly employed, even
          though he may have been temporarily working in another department at the time of injury.)

THE ACCIDENT OR EXPOSURE TO OCCUPATIONAL ILLNESS

  10. Place of accident or exposure _____
          (No. and street)   (City or town)   (State)
      If accident or exposure occurred on employer's premises, give address of plant or establishment in which
      it occurred. Do not indicate department or division within the plant or establishment. If accident oc-
      curred outside employer's premises at an identifiable address, give that address. If it occurred on a pub-
      lic highway or at any other place which cannot be identified by number and street, please provide place
      references locating the place of injury as accurately as possible.

  11. Was place of accident or exposure on employer's premises? _____± (Yes or No)

  12. What was the employee doing when injured? _____
                 (Be specific. If he was using tools or equipment or handling material,
_____
                name them and tell what he was doing with them.)

  13. How did the accident occur? _____
              (Describe fully the events which resulted in the injury or occupational illness. Tell what
_____
happened and how it happened. Name any objects or substances involved and tell how they were involved. Give
_____
full details on all factors which led or contributed to the accident. Use separate sheet for additional space.)

OCCUPATIONAL INJURY OR OCCUPATIONAL ILLNESS

  14. Describe the injury or illness in detail and indicate the part of body affected. _____
_____
                    (e.g.: amputation of right index finger
          at second joint; fracture of ribs; lead poisoning; dermatitis of left hand, etc.)

  15. Name the object or substance which directly injured the employee. (For example, the machine or thing
      he struck against or which struck him; the vapor or poison he inhaled or swallowed; the chemical or ra-
      diation which irritated his skin; or in cases of strains, hernias, etc., the thing he was lifting, pulling, etc.)
_____
_____

  16. Date of injury or initial diagnosis of occupational illness _____
                    (Date)

  17. Did employee die? _____ (Yes or No)

OTHER

  18. Name and address of physician _____

  19. If hospitalized, name and address of hospital _____
_____
      Date of report _____ Prepared by _____
      Official position _____

CLASSIFY THE HAZARDS

Hazard Classification

- Class "A" Hazard — Any condition or practice with potential for causing loss of life or body part and/or extensive loss of structure, equipment, or material.
- Class "B" Hazard — Any condition or practice with potential for causing serious injury, illness, or property damage but less severe than Class "A".
- Class "C" Hazard — Any condition or practice with probable potential for causing nondisabling injury or nondisabling illness, or nondisruptive property damage.

Loss Severity Potential

Using the sample Safety And Health Self-Inspection Reports on the next following pages, prioritize the six listed hazards as follows,

Please note that the No. 6 hazard is the only hazard that has the potential to cause loss of life or body part, etc. Therefore it would be first in the order of abatement priority.

The next hazard in the order of priority would be No. 1 and No. 4 — both are class "B" hazards.

The last two hazards are classified as class "C" hazards.

Classifying your inspected hazards allows your maintenance personnel to concentrate their efforts on the most serious hazards, and then on down to the least serious hazards.

The consequence of not setting individual hazard priorities could expose any one of your workers to an extremely hazardous condition that could result in loss of life — while your maintenance personnel are abating a less serious hazard.

PLANNED SAFETY AND HEALTH SELF-INSPECTION GUIDE

DEPARTMENT Maintenance	UNIT Workshop	SUPERVISOR RESPONSIBLE J. P. Smith	APPROVED BY Ralph T. Welles	DATE 4/16/72	PAGE NO. 1
1. PROBLEMS	2. CRITICAL FACTORS	3. CONDITIONS TO OBSERVE	4. FREQUENCY	5. RESPONSIBILITY	
1. Overhead hoist	Cables, chains, hooks, pulleys	Frayed or deformed cables, worn or broken hooks and chains, damaged pulleys	Daily – before each shift	Operators	
2. Hydraulic pump	High pressure hose	Leaks; broken or loose fittings	Daily	Shift leader	
3. Power generator	High voltage lines	Frayed or broken insulation	Weekly	Foreman	
4. Fire extinguishers	Contents, location, charge	Correct type, fully charged, properly located, corrosion, leaks	Monthly	Area safety inspector	
5. General housekeeping	Passageways, aisles, floors, grounds	Free of obstructions, clearly marked, free of refuse	Daily	Shift leader foreman	

PLANNED SAFETY AND HEALTH SELF-INSPECTION GUIDE

DEPARTMENT	UNIT	SUPERVISOR RESPONSIBLE	APPROVED BY	DATE	PAGE NO.

1. PROBLEMS	2. CRITICAL FACTORS	3. CONDITIONS TO OBSERVE	4. FREQUENCY	5. RESPONSIBILITY

SAFETY AND HEALTH SELF-INSPECTION REPORT

Area Inspected _____ _Bldg. A_ _____ Date _7/30/72_ Inspector _John Williamson_

CODE	HAZARDS	CORRECTIVE ACTIONS
*①B	Guard missing on shear blade, #2 machine, S.W. corner of Bay #1 (7/16/72)	W.O. issued to Eng. for new guard (7/16/72) - wooden barrier guard in temporary use - guard still missing 7/30/72 (contacted Eng. and they will install guard by 8/3/72)
*2C	Window still cracked on S. wall, Bay #3 - W.O. was issued for replacement on 7/16/72	Maint. Dept. now plans to replace all broken windows in plant starting Aug. 3
⊗	Oil and trash accumulated under main motor in pump room. Was to be cleaned out by 7/31/72	Cleaned out 7/31/72. Employees instructed to keep area clean, and why.
*④B	Mirror at pedestrian walk, corner of N. end of mach. shop, out of line	Jack Butler scheduled adjust- ment for Aug. 8 - temporary warning sign to be posted 7/31/72
✗	Three employees at cleaning tank in electric shop not wearing eye protection	Discussed with Roberts - he held meeting with his employees on eye protection rules and benefits
⑥A	Cable on jib crane, Bay #3, badly frayed	Called Don Johnson who tagged crane out of service. Cable will be replaced Sat., Aug. 1

ANALYSIS AND COMMENTS

    Class "A" Hazards now getting good, high priority attention. Steady

    progress being made on others, too.

SAFETY AND HEALTH SELF-INSPECTION REPORT

Area Inspected_____ Date_____ Inspector_____

CODE	HAZARDS	CORRECTIVE ACTIONS

ANALYSIS AND COMMENTS

Summary

Occupational Injuries and Illnesses

Establishment Name and Address:

Code 1	Category 2	Fatalities 3	Lost Workday Cases			Nonfatal Cases Without Lost Workdays*	
	Injury and Illness Category		Number of Cases 4	Number of Cases Involving Permanent Transfer to Another Job or Termination of Employment 5	Number of Lost Workdays 6	Number of Cases 7	Number of Cases Involving Transfer to Another Job or Termination of Employment 8
10	Occupational Injuries	0	1	0	1	0	0
	Occupational Illnesses						
21	Occupational Skin Diseases or Disorders	0	0	0	0	0	0
22	Dust diseases of the lungs (pneumoconioses)	0	0	0	0	0	0
23	Respiratory conditions due to toxic agents	0	0	0	0	0	0
24	Poisoning (systemic effects of toxic materials)	0	1	1	4	0	0
25	Disorders due to physical agents (other than toxic materials)	0	0	0	0	0	0
26	Disorders due to repeated trauma	0	0	0	0	0	0
29	All other occupational illnesses	0	0	0	0	0	0
	Total—occupational illnesses (21-29)	0	1	1	4	0	0
	Total—occupational injuries and illnesses	0	2	1	5	0	0

*Nonfatal Cases Without Lost Workdays—Cases resulting in: Medical treatment beyond first aid, diagnosis of occupational illness, loss of consciousness, restriction of work or motion, or transfer to another job (without lost workdays).

Bureau of Labor Statistics
Supplementary Record of
Occupational Injuries and Illnesses

U.S. Department of Labor

Form Approved
O.M.B. No. 1220 0029

This form is required by Public Law 91-596 and must be kept in the establishment for 5 years. | Case or File No.
Failure to maintain can result in the issuance of citations and assessment of penalties.

Employer

1. Name

2. Mail address *(No. and street, city or town, State, and zip code)*

3. Location, if different from mail address

Injured or Ill Employee

4. Name *(First, middle, and last)*   Social Security No.

5. Home address *(No. and street, city or town, State, and zip code)*

6. Age   7. Sex: *(Check one)*   Male ☐   Female ☐

8. Occupation *(Enter regular job title, not the specific activity he was performing at time of injury.)*

9. Department *(Enter name of department or division in which the injured person is regularly employed, even though he may have been temporarily working in another department at the time of injury.)*

The Accident or Exposure to Occupational Illness

If accident or exposure occurred on employer's premises, give address of plant or establishment in which it occurred. Do not indicate department or division within the plant or establishment. If accident occurred outside employer's premises at an identifiable address, give that address. If it occurred on a public highway or at any other place which cannot be identified by number and street, please provide place references locating the place of injury as accurately as possible.

10. Place of accident or exposure *(No. and street, city or town, State, and zip code)*

11. Was place of accident or exposure on employer's premises?   Yes ☐   No ☐

12. What was the employee doing when injured? *(Be specific. If he was using tools or equipment or handling material, name them and tell what he was doing with them.)*

13. How did the accident occur? *(Describe fully the events which resulted in the injury or occupational illness. Tell what happened and how it happened. Name any objects or substances involved and tell how they were involved. Give full details on all factors which led or contributed to the accident. Use separate sheet for additional space.)*

Occupational Injury or Occupational Illness

14. Describe the injury or illness in detail and indicate the part of body affected. *(E.g., amputation of right index finger at second joint; fracture of ribs; lead poisoning; dermatitis of left hand, etc.)*

15. Name the object or substance which directly injured the employee. *(For example, the machine or thing he struck against or which struck him; the vapor or poison he inhaled or swallowed; the chemical or radiation which irriatated his skin; or in cases of strains, hernias, etc., the thing he was lifting, pulling, etc.)*

16. Date of injury or initial diagnosis of occupational illness

17. Did employee die? *(Check one)* Yes ☐ No ☐

Other

18. Name and address of physician

19. If hospitalized, name and address of hospital

Date of report	Prepared by		Official position

OSHA No. 101 (Feb. 1981)

SUPPLEMENTARY RECORD OF OCCUPATIONAL INJURIES AND ILLNESSES

To supplement the Log and Summary of Occupational Injuries and Illnesses (OSHA No. 200), each establishment must maintain a record of each recordable occupational injury or illness. Worker's compensation, insurance, or other reports are acceptable as records if they contain all facts listed below or are supplemented to do so. If no suitable report is made for other purposes, this form (OSHA No. 101) may be used or the necessary facts can be listed on a separate plain sheet of paper. These records must also be available in the establishment without delay and at reasonable times for examination by representatives of the Department of Labor and the Department of Health and Human Services, and States accorded jurisdiction under the Act. The records must be maintained for a period of not less than five years following the end of the calendar year to which they relate.

Such records must contain at least the following facts:

1) *About the employer*—name, mail address, and location if different from mail address.

2) *About the injured or ill employee*—name, social security number, home address, age, sex, occupation, and department.

3) *About the accident or exposure to occupational illness*—place of accident or exposure, whether it was on employer's premises, what the employee was doing when injured, and how the accident occurred.

4) *About the occupational injury or illness*—description of the injury or illness, including part of body affected; name of the object or substance which directly injured the employee; and date of injury or diagnosis of illness.

5) *Other*—name and address of physician; if hospitalized, name and address of hospital; date of report; and name and position of person preparing the report.

SEE *DEFINITIONS* ON THE BACK OF OSHA FORM 200.

SUPPLEMENTARY RECORD OF OCCUPATIONAL INJURIES AND ILLNESSES

To supplement the Log and Summary of Occupational Injuries and Illnesses (OSHA No. 200), each establishment must maintain a record of each recordable occupational injury or illness. Worker's compensation, insurance, or other reports are acceptable as records if they contain all facts listed below or are supplemented to do so. If no suitable report is made for other purposes, this form (OSHA No. 101) may be used or the necessary facts can be listed on a separate plain sheet of paper. These records must also be available in the establishment without delay and at reasonable times for examination by representatives of the Department of Labor and the Department of Health and Human Services, and States accorded jurisdiction under the Act. The records must be maintained for a period of not less than five years following the end of the calendar year to which they relate.

Such records must contain at least the following facts:

1) About the employer — name, mail address, and location if different from mail address.
2) About the injured or ill employee — name, social security number, home address, age, sex, occupation, and department.
3) About the accident or exposure to occupational illness — place of accident or exposure, whether it was on employer's premises, what the employee was doing when injured, and how the accident occurred.
4) About the occupational injury or illness — description of the injury or illness, including part of body affected; name of the object or substance which directly injured the employee; and date of injury or diagnosis of illness.
5) Other — name and address of physician; if hospitalized, name and address of hospital; date of report; and name and position of person preparing the report.

SEE *DEFINITIONS* ON THE BACK OF OSHA FORM 200.

Sample Job Hazard Analysis
Cleaning Inside Surface of Chemical Tank—Top Manhole Entry

STEP	HAZARD	NEW PROCEDURE OR PROTECTION
1. Select and train operators.	Operator with respiratory or heart problem; other physical limitation.	• Examination by industrial physician for suitability to work.
	Untrained operator—failure to perform task.	• Train operators. • Dry run. [Reference: National Institute for Occupational Safety and Health (NIOSH) Doc. #80-406]
2. Determine what is in the tank, what process is going on in the tank, and what hazards this can pose.	Explosive gas.	• Obtain work permit signed by safety, maintenance and supervisors.
	Improper oxygen level.	• Test air by qualified person.
	Chemical exposure— Gas, dust, vapor: irritant toxic Liquid: irritant toxic corrosive: Solid: irritant corrosive	• Ventilate to 19.5%-23.5% oxygen and less than 10% LEL of any flammable gas. Steaming inside of tank, flushing and draining, then ventilating, as previously described, may be required. • Provide appropriate respiratory equipment —SCBA or air line respirator. • Provide protective clothing for head, eyes, body and feet. • Provide parachute harness and lifeline. [Reference: OSHA standards 1910.106, 1926.100, 1926.21(b)(6); NIOSH Doc. #80-406] • Tanks should be cleaned from outside, if possible.
3. Set up equipment.	Hoses, cord, equipment — tripping hazards.	• Arrange hoses, cords, lines and equipment in orderly fashion, with room to maneuver safely.
	Electrical — voltage too high, exposed conductors.	• Use ground-fault circuit interrupter.
	Motors not locked out and tagged.	• Lockout and tag mixing motor, if present.

	Task	Hazard	Recommended Action
4	Install ladder in tank.	Ladder slipping.	• Secure to manhole top or rigid structure.
5	Prepare to enter tank.	Gas or liquid in tank.	• Empty tank through existing piping. • Review emergency procedures. • Open tank. • Check of job site by industrial hygienist or safety professional. • Install blanks in flanges in piping to tank. (Isolate tank.) • Test atmosphere in tank by qualified person (long probe).
6	Place equipment at tank-entry position.	Trip or fall.	• Use mechanical-handling equipment. • Provide guardrails around work positions at tank top.
7	Enter tank.	Ladder — tripping hazard.	• Provide personal protective equipment for conditions found. [Reference: NIOSH Doc. #80-406, OSHA CFR 1910.134]
		Exposure to hazardous atmosphere.	• Provide outside helper to watch, instruct and guide operator entering tank, with capability to lift operator from tank in emergency.
8	Cleaning tank.	Reaction of chemicals, causing mist or expulsion of air contaminant.	• Provide protective clothing and equipment for all operators and helpers. • Provide lighting for tank (Class I, Div. 1). • Provide exhaust ventilation. • Provide air supply to interior of tank. • Frequent monitoring of air in tank. • Replace operator or provide rest periods. • Provide means of communication to get help, if needed. • Provide two-man standby for any emergency.
9	Cleaning up.	Handling of equipment, causing injury.	• Dry run. • Use material-handling equipment.

SUPERVISOR'S INVESTIGATION REPORT

Company _____   Department _____

Exact location
of event _____   Date and time
of event _____

EMPLOYEE IDENTIFICATION

Name _____   Occupation _____

Nature of injury or illness _____

Object/equipment/substance which
inflicted injury or caused illness _____

Person with most control of
object/equipment/substance _____

DESCRIPTION OF EVENT

____ (Describe in detail what occurred and how the event occurred) ____

_____

_____

ANALYSIS OF CAUSES

Immediate causes   (What hazardous acts and/or hazardous conditions

_____ contributed most directly to the event) _____

Basic causes _____ (What are the basic causes for the existence

_____ of these hazardous acts or conditions) _____

_____

PREVENTIVE ACTION

____ (What action has, or will be taken to prevent recurrence) _____

_____

LOSS SEVERITY POTENTIAL			PROBABLE RECURRENCE RATE		
Major	Serious	Minor	Frequent	Occasional	Rare
____	____	____	____	____	____

Investigated by _____   Date _____

Job Hazard Analysis Form

JOB TITLE: DATE OF ANALYSIS:

JOB LOCATION:

STEP	HAZARD	NEW PROCEDURE OR PROTECTION

Material Safety Data Sheet

May be used to comply with
OSHA's Hazard Communication Standard,
29 CFR 1910.1200. Standard must be
consulted for specific requirements.

U.S. Department of Labor

Occupational Safety and Health Administration
(Non-Mandatory Form)
Form Approved
OMB No. 1218-0072

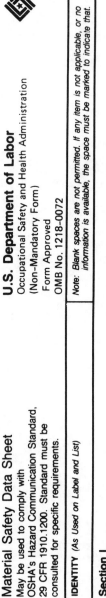

IDENTITY (As Used on Label and List)	Note: Blank spaces are not permitted. If any item is not applicable, or no information is available, the space must be marked to indicate that.

Section I

Manufacturer's Name	Emergency Telephone Number
Address (Number, Street, City, State, and ZIP Code)	Telephone Number for Information
	Date Prepared
	Signature of Preparer (optional)

Section II — Hazardous Ingredients/Identity Information

Hazardous Components (Specific Chemical Identity; Common Name(s))	OSHA PEL	ACGIH TLV	Other Limits Recommended	% (optional)

Section III — Physical/Chemical Characteristics

Boiling Point	Specific Gravity (H$_2$O = 1)
Vapor Pressure (mm Hg.)	Melting Point
Vapor Density (AIR = 1)	Evaporation Rate (Butyl Acetate = 1)
Solubility in Water	
Appearance and Odor	

Section IV — Fire and Explosion Hazard Data

Flash Point (Method Used)	Flammable Limits	LEL	UEL

Extinguishing Media

Special Fire Fighting Procedures

Unusual Fire and Explosion Hazards

(Reproduce locally)   OSHA 174, Sept. 1985

Section V — Reactivity Data

Stability	Unstable	Conditions to Avoid
	Stable	

Incompatibility (Materials to Avoid)

Hazardous Decomposition or Byproducts

Hazardous Polymerization	May Occur	Conditions to Avoid
	Will Not Occur	

Section VI — Health Hazard Data

Route(s) of Entry:	Inhalation?	Skin?	Ingestion?

Health Hazards (Acute and Chronic)

Carcinogenicity:	NTP?	IARC Monographs?	OSHA Regulated?

Signs and Symptoms of Exposure

Medical Conditions
Generally Aggravated by Exposure

Emergency and First Aid Procedures

Section VII — Precautions for Safe Handling and Use

Steps to Be Taken in Case Material Is Released or Spilled

Waste Disposal Method

Precautions to Be Taken in Handling and Storing

Other Precautions

Section VIII — Control Measures

Respiratory Protection (Specify Type)

Ventilation	Local Exhaust		Special
	Mechanical (General)		Other
Protective Gloves		Eye Protection	

Other Protective Clothing or Equipment

Work/Hygienic Practices

Page 2

☆ U S G P O 1986 - 491 - 529 / 45775

EXHIBIT 14-2

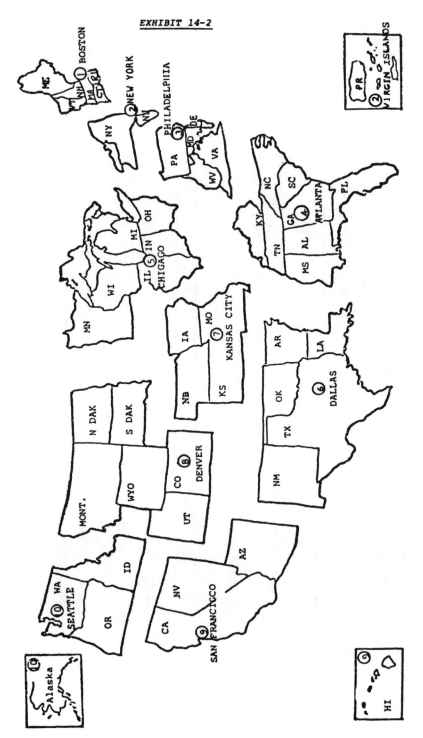

Regional offices U.S. Department of Labor, Occupational Safety and Health Administration.

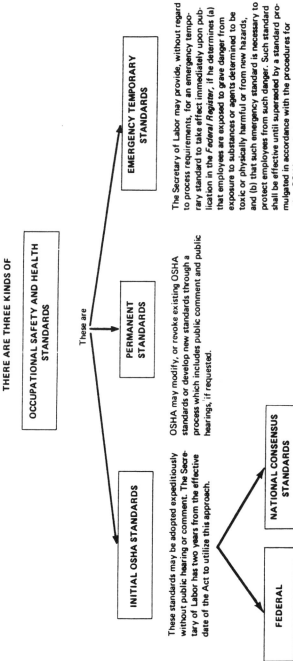

THERE ARE THREE KINDS OF

OCCUPATIONAL SAFETY AND HEALTH STANDARDS

These are

INITIAL OSHA STANDARDS

These standards may be adopted expeditiously without public hearing or comment. The Secretary of Labor has two years from the effective date of the Act to utilize this approach.

FEDERAL

These standards come from existing Federal Laws such as:

The Walsh-Healy Act
The Construction Safety Act
The Service Contract Act
Longshoremen and Harbor Workers Act
The National Arts and Humanities Act

NATIONAL CONSENSUS STANDARDS

These standards come from industrial associations such as:

The American Standards Institute (ANSI)
The National Fire Protection Association (NFPA)

PERMANENT STANDARDS

OSHA may modify, or revoke existing OSHA standards or develop new standards through a process which includes public comment and public hearings, if requested.

EMERGENCY TEMPORARY STANDARDS

The Secretary of Labor may provide, without regard to process requirements, for an emergency temporary standard to take effect immediately upon publication in the *Federal Register*, if he determines (a) that employees are exposed to grave danger from exposure to substances or agents determined to be toxic or physically harmful or from new hazards, and (b) that such emergency standard is necessary to protect employees from such danger. Such standard shall be effective until superseded by a standard promulgated in accordance with the procedures for other OSHA standards.

DEFINITIONS

abate — to correct under the Occupational Safety and Health Act; to come into compliance with a standard that is being violated.

comply — to act in accordance with the Occupational Safety and Health Standards; to follow the rules and regulations published in the CODE OF FEDERAL REGULATIONS.

contest — to object to or to appeal a decision made by the OSHA Area Director.

citation — issued by the OSHA Area Director to an employer for an alleged violation reported by the OSHA Compliance and Safety and Health Officer during a compliance visit.

"de minimis" violation — violation of a standard that does not involve an immediate or direct relationship to the safety or health of an employee.

hazard — a risk, danger, or peril to employees in the workplace.

imminent danger — any condition or practice in any place of employment which is such that a danger exists which could reasonably be expected to cause death or serious physical harm immediately or before the imminence of such danger can be eliminated through the enforcement procedures otherwise provided by the Act.

promulgate — to issue, establish, or make known officially the terms of a law or regulation having the force of law. For example, to publish OSHA rules, procedures, standards, and regulations in the FEDERAL REGISTER.

"recognized" hazard — a hazard is recognized if it is a condition that is (a) of common knowledge or general recognition in the particular industry in which it occurs, and (b) detectable (1) by means of senses (sight, smell, touch, and hearing), or (2) is of such wide, general recognition as a hazard in the industry that even if it is not detectable by means of the senses, there are generally known and accepted tests for its existence which should make its presence known to the employer. For example, excessive concentrations of a toxic substance in the air would be a "recognized" hazard even though they could be detected only through the use of measuring devices.

repeated violation — a violation of any standard, rule or order, or the general duty clause, may be cited as repeated under the Act (Section 17a) where a second citation is issued under the Act for violation of the same standard, rule, or order, or the same condition violating the general duty clause for which a previous citation was issued. A repeated violation differs from a failure to abate in that repeated violations exist where the employer has abated an earlier violation and, upon later inspection, is found to have violated the same standard. A notice of failure to abate would be appropriate where the employer has been cited and fails to abate the hazard cited within the abatement period.

standard — a rule, established in accordance with law or other competent authority, which designates safe and healthful conditions or practices by which work must be performed to prevent injury or illness.

variance — formal approval by OSHA permitting an employer to bypass certain requirements of the standards. A temporary variance may be granted if the employer can show he is unable to comply by its effective date due to lack of personnel, materials, equipment, or because alterations or construction required for compliance cannot be accomplished within the specified time. A permanent variance may be granted if he can prove that he is providing safe and healthful working conditions equal to those which would pertain if he had complied. Variances may also be granted if the employer is participating in an approved worker safety and health experiment, or if compliance would constitute a serious impairment of national defense.

Concepts and Techniques of Machine Safeguarding

U.S. Department of Labor
Raymond J. Donovan, Secretary

Occupational Safety and Health Administration
Thorne G. Auchter, Assistant Secretary

For sale by the Superintendent of Documents, U.S. Government Printing Office
Washington, D.C. 20402

Introduction

This manual has been prepared as an aid to employers, employees, machine manufacturers, machine guard designers and fabricators, and all others with an interest in protecting workers against the hazards of moving machine parts. It identifies the major mechanical motions and the general principles of safeguarding them. Current applications of each technique are shown in accompanying illustrations of specific operations and machines. The concepts described here may be transferred, with due care, to different machines with similar motions. Whether or not a proper safeguard has been manufactured for a particular application, no mechanical motion that threatens a worker's safety should be left without a safeguard.

All possible approaches to machine safeguarding are not discussed in this manual. Why? Because practical solutions to moving machine parts problems are as numerous as the people working on them. No publication could keep pace with reports of these solutions or attempt to depict them all.

In machine safeguarding, as in other regulated areas of the American workplace, to a certain extent OSHA standards govern function and practice. This text, however, is not a substitute for the standards. It is a manual of basic technical information and workable ideas which the employer may use as a guide to voluntary compliance. It offers an overview of the machine safeguarding problem in its industrial setting, an assortment of solutions in popular use, and a challenge to all whose work involves machines.

Many readers of this manual already have the judgment, knowledge, and skill to develop effective answers to problems yet unsolved. Innovators are encouraged to find here stimulation to eliminate mechanical hazards facing America's workers today.

Chapter 1

Basics of Machine Safeguarding

Crushed hands and arms, severed fingers, blindness—the list of possible machinery-related injuries is as long as it is horrifying. There seem to be as many hazards created by moving machine parts as there are types of machines. Safeguards are essential for protecting workers from needless and preventable injuries.

A good rule to remember is: Any machine part, function, or process which may cause injury must be safeguarded. Where the operation of a machine or accidental contact with it can injure the operator or others in the vicinity, the hazard must be either controlled or eliminated.

This manual describes the various hazards of mechanical motion and action and presents some techniques for protecting workers from these hazards. General information is covered in this chapter—where mechanical hazards occur, what kinds of motions need safeguarding, and what the requirements are for effective safeguards, as well as a brief discussion of nonmechanical hazards and some other considerations.

Where Mechanical Hazards Occur

Dangerous moving parts in these three basic areas need safeguarding:

The point of operation: that point where work is performed on the material, such as cutting, shaping, boring, or forming of stock.

Power transmission apparatus: all components of the mechanical system which transmit energy to the part of the machine performing the work. These components include flywheels, pulleys, belts, connecting rods, couplings, cams, spindles, chains, cranks, and gears.

Other moving parts: all parts of the machine which move while the machine is working. These can include reciprocating, rotating, and transverse moving parts, as well as feed mechanisms and auxiliary parts of the machine.

Hazardous Mechanical Motions and Actions

A wide variety of mechanical motions and actions may present hazards to the worker. These can include the movement of rotating members, reciprocating arms, moving belts, meshing gears, cutting teeth, and any parts that impact or shear. These different types of hazardous mechanical motions and actions are basic to nearly all machines, and recognizing them is the first step toward protecting workers from the danger they present.

The basic types of hazardous mechanical motions and actions are:

Motions
- rotating (including in-running nip points)
- reciprocating
- transverse

Actions
- cutting
- punching
- shearing
- bending

We will briefly examine each of these basic types in turn.

Motions

Rotating motion can be dangerous; even smooth, slowly rotating shafts can grip clothing, and through mere skin contact force an arm or hand into a dangerous position. Injuries due to contact with rotating parts can be severe.

Collars, couplings, cams, clutches, flywheels, shaft ends, spindles, and horizontal or vertical shafting are some examples of common rotating mechanisms which may be hazardous. The danger increases when bolts, nicks, abrasions, and projecting keys or set screws are exposed on rotating parts, as shown in Figure 1.

In-running nip point hazards are caused by the rotating parts on machinery. There are three main types of in-running nips.

Parts can rotate in opposite directions while their axes are

Figure 1.

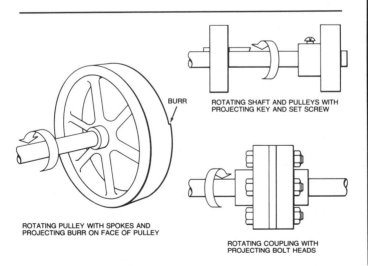

BURR

ROTATING SHAFT AND PULLEYS WITH
PROJECTING KEY AND SET SCREW

ROTATING PULLEY WITH SPOKES AND
PROJECTING BURR ON FACE OF PULLEY

ROTATING COUPLING WITH
PROJECTING BOLT HEADS

parallel to each other. These parts may be in contact (producing a nip point) or in close proximity to each other. In the latter case the *stock* fed between the rolls produces the nip points. This danger is common on machinery with intermeshing gears, rolling mills, and calenders. See Figure 2.

Figure 2.

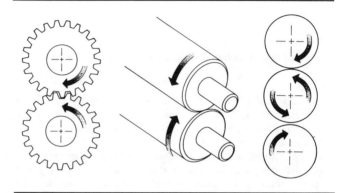

Another nip point is created between rotating and tangentially moving parts. Some examples would be: the point of contact between a power transmission belt and its pulley, a chain and a sprocket, or a rack and pinion. See Figure 3.

Nip points can occur between rotating and fixed parts which create a shearing, crushing, or abrading action. Examples are: spoked handwheels or flywheels, screw conveyors, or the periphery of an abrasive wheel and an incorrectly adjusted work rest. See Figure 4.

Reciprocating motions may be hazardous because, during the back-and-forth or up-and-down motion, a worker may be struck by or caught between a moving and a stationary part. See Figure 5 for an example of a reciprocating motion.

Transverse motion (movement in a straight, continuous line) creates a hazard because a worker may be struck or caught in a pinch or shear point by the moving part. See Figure 6.

Actions

Cutting action involves rotating, reciprocating, or transverse motion. The danger of cutting action exists at the point of operation where finger, head, and arm injuries can occur and where flying chips or scrap material can strike the eyes or face. Such hazards are present at the point of operation in cutting wood, metal, or other materials. Typical examples of mechanisms in-

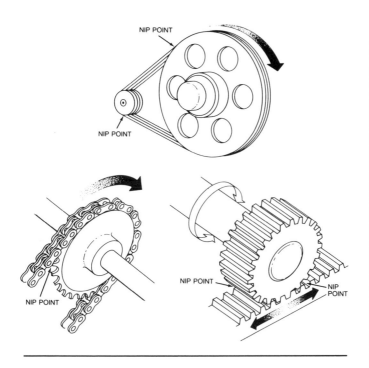

NIP POINT

NIP POINT

NIP POINT

NIP POINT

NIP POINT

NIP POINT

Figure 3.

volving cutting hazards include bandsaws, circular saws, boring or drilling machines, turning machines (lathes), or milling machines. See Figure 7.

Punching action results when power is applied to a slide (ram) for the purpose of blanking, drawing, or stamping metal or other materials. The danger of this type of action occurs at the point of operation where stock is inserted, held, and withdrawn by hand.

Typical machinery used for punching operations are power presses and iron workers. See Figure 8.

Shearing action involves applying power to a slide or knife in order to trim or shear metal or other materials. A hazard occurs at the point of operation where stock is actually inserted, held, and withdrawn.

Typical examples of machinery used for shearing operations are mechanically, hydraulically, or pneumatically powered shears. See Figure 9.

Bending action results when power is applied to a slide in order

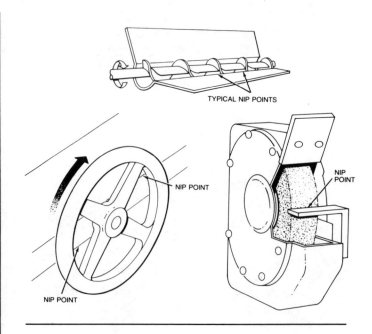

Figure 4.

Figure 5.

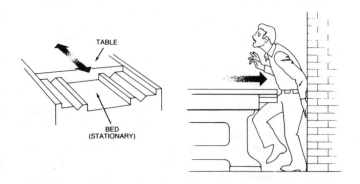

TRANSVERSE MOTION OF BELT

Figure 6.

Figure 7.

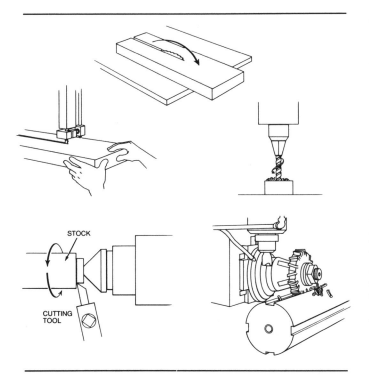

STOCK

CUTTING
TOOL

to draw or stamp metal or other materials, and a hazard occurs at the point of operation where stock is inserted, held, and withdrawn.

Equipment that uses bending action includes power presses, press brakes, and tubing benders. See Figure 10.

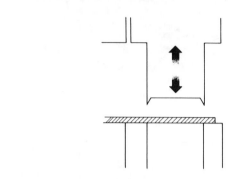

Figure 8.

Requirements for Safeguards

What must a safeguard do to protect workers against mechanical hazards? Safeguards must meet these minimum general requirements:

Prevent contact: The safeguard must prevent hands, arms, or any other part of a worker's body from making contact with dangerous moving parts. A good safeguarding system eliminates the possibility of the operator or another worker placing their hands near hazardous moving parts.

Figure 9.
Shearing

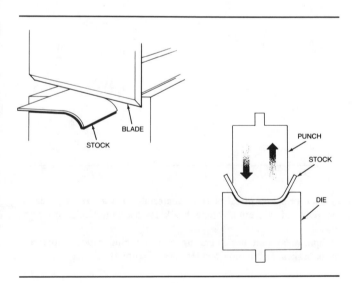

Figure 10.
Bending

Secure: Workers should not be able to easily remove or tamper with the safeguard, because a safeguard that can easily be made ineffective is no safeguard at all. Guards and safety devices should be made of durable material that will withstand the conditions of normal use. They must be firmly secured to the machine.

Protect from falling objects: The safeguard should ensure that no objects can fall into moving parts. A small tool which is dropped into a cycling machine could easily become a projectile that could strike and injure someone.

Create no new hazards: A safeguard defeats its own purpose if it creates a hazard of its own such as a shear point, a jagged edge, or an unfinished surface which can cause a laceration. The edges of guards, for instance, should be rolled or bolted in such a way that they eliminate sharp edges.

Create no interference: Any safeguard which impedes a worker from performing the job quickly and comfortably might soon be overridden or disregarded. Proper safeguarding can actually enhance efficiency since it can relieve the worker's apprehensions about injury.

Allow safe lubrication: If possible, one should be able to lubricate the machine without removing the safeguards. Locating oil reservoirs outside the guard, with a line leading to the lubrication point, will reduce the need for the operator or maintenance worker to enter the hazardous area.

Nonmechanical Hazards

While this manual concentrates attention on concepts and techniques for safeguarding mechanical motion, machines obviously present a variety of other hazards which cannot be ignored. Full discussion of these matters is beyond the scope of this publication, but some nonmechanical hazards are briefly mentioned below to remind the reader of things other than safeguarding moving parts which can affect the safe operation of machinery.

All power sources for machinery are potential sources of danger. When using electrically powered or controlled machines, for instance, the equipment as well as the electrical system itself must be properly grounded. Replacing frayed, exposed, or old wiring will also help to protect the operator and others from electrical shocks or electrocution. High pressure systems, too, need careful inspection and maintenance to prevent possible failure from pulsation, vibration, or leaks. Such a failure could cause explosions or flying objects.

Machines often produce noise (unwanted sound) and this can result in a number of hazards to workers. Not only can it startle and disrupt concentration, but it can interfere with communications, thus hindering the worker's safe job performance. Research has linked noise to a whole range of harmful health effects, from hearing loss and aural pain to nausea, fatigue, reduced muscle control, and emotional disturbances. Engineering controls such as the use of sound-dampening materials, as well as less sophisticated hearing protection, such as ear plugs and muffs, have been suggested as ways of controlling the harmful effects of noise. Vibration, a related hazard which can cause noise and thus result in fatigue and illness for the worker, may be avoided if machines are properly aligned, supported, and, if necessary, anchored.

Because some machines require the use of cutting fluids, coolants, and other potentially harmful substances, operators, maintenance workers, and others in the vicinity may need protection. These substances can cause ailments ranging from dermatitis to serious illnesses and disease. Specially constructed safeguards, ventilation, and protective equipment and clothing are possible temporary solutions to the problem of machinery-related chemical hazards until these hazards can be better controlled or eliminated from the workplace.

Training

Even the most elaborate safeguarding system cannot offer effective protection unless the worker knows how to use it and why. Specific and detailed training is therefore a crucial part of any effort to provide safeguarding against machine-related hazards. Thorough operator training should involve instruction or hands-on training in the following:

(1) a description and identification of the hazards associated with particular machines;

(2) the safeguards themselves, how they provide protection, and the hazards for which they are intended;

(3) how to use the safeguards and why;

(4) how and under what circumstances safeguards can be removed, and by whom (in most cases, repair or maintenance personnel only); and

(5) what to do (e.g., contact the supervisor) if a safeguard is damaged, missing, or unable to provide adequate protection.

This kind of safety training is necessary for new operators and maintenance or setup personnel, when any new or altered safeguards are put in service, or when workers are assigned to a new machine or operation.

Protective Clothing and Personal Protective Equipment

Engineering controls, which eliminate the hazard at the source and do not rely on the worker's behavior for their effectiveness, offer the best and most reliable means of safeguarding. Therefore, engineering controls must be the employer's first choice for eliminating machinery hazards. But whenever an extra measure of protection is necessary, operators must wear protective clothing or personal protective equipment.

If it is to provide adequate protection, the protective clothing and equipment selected must always be:

(1) appropriate for the particular hazards;
(2) maintained in good condition;
(3) properly stored when not in use, to prevent damage or loss; and
(4) kept clean and sanitary.

Protective clothing is, of course, available for different parts of the body. Hard hats can protect the head from the impact of bumps and falling objects when the worker is handling stock; caps and hair nets can help keep the worker's hair from being caught in machinery. If machine coolants could splash or particles could fly into the operator's eyes or face, then face shields, safety goggles, glasses, or similar kinds of protection might be necessary. Hearing protection may be needed when workers operate noisy machinery. To guard the trunk of the body from cuts or impacts from heavy or rough-edged stock, there are certain protective coveralls, jackets, vests, aprons, and full-body suits. Workers can protect their hands and arms from the same kinds of injury with special sleeves and gloves. And safety shoes and boots, or other acceptable foot guards, can shield the feet against injury in case the worker needs to handle heavy stock which might drop.

It is important to note that protective clothing and equipment themselves can create hazards. A protective glove which can become caught between rotating parts, or a respirator facepiece which hinders the wearer's vision, for example, require alertness and careful supervision whenever they are used.

Other aspects of the worker's dress may present additional safety hazards. Loose-fitting clothing might possibly become entangled in rotating spindles or other kinds of moving machinery. Jewelry, such as bracelets and rings, can catch on machine parts or stock and lead to serious injury by pulling a hand into the danger area.

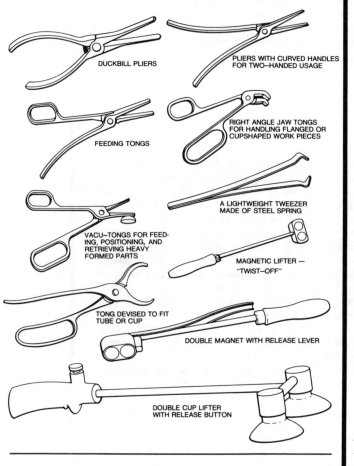

DUCKBILL PLIERS

PLIERS WITH CURVED HANDLES
FOR TWO—HANDED USAGE

FEEDING TONGS

RIGHT ANGLE JAW TONGS
FOR HANDLING FLANGED OR
CUPSHAPED WORK PIECES

A LIGHTWEIGHT TWEEZER
MADE OF STEEL SPRING

VACU—TONGS FOR FEED-
ING, POSITIONING, AND
RETRIEVING HEAVY
FORMED PARTS

MAGNETIC LIFTER —
"TWIST—OFF"

TONG DEVISED TO FIT
TUBE OR CUP

DOUBLE MAGNET WITH RELEASE LEVER

DOUBLE CUP LIFTER
WITH RELEASE BUTTON

Figure 66.
Holding tools

Figure 67.

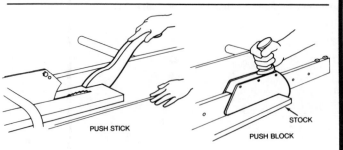

PUSH STICK

STOCK

PUSH BLOCK

Chapter 3 | # Guard Construction

Today many builders of single-purpose machines provide point-of-operation and power transmission safeguards as standard equipment. However, not all machines in use have built-in safeguards provided by the manufacturer.

Guards designed and installed by the *builder* offer two main advantages:

- They usually conform to the design and function of the machine.

- They can be designed to strengthen the machine in some way or to serve some additional functional purposes.

User-built guards are sometimes necessary for a variety of reasons. They have these advantages:

- Often, with older machinery, they are the only practical solution.

- They may be the only choice for mechanical power transmission apparatus in older plants, where machinery is not powered by individual motor drive.

- They permit options for point-of-operation safeguards when skilled personnel and machinery are available to make them.

- They can be designed and built to fit unique and even changing situations.

- They can be installed on individual dies and feeding mechanisms.

- Design and installation of machine safeguards by plant personnel can help to promote safety consciousness in the workplace.

However, they also have disadvantages:

- User-built guards may not conform well to the configuration and function of the machine.

- There is a risk that user-built guards may be poorly designed or built.

Point-of-Operation Guards

Point-of-operation guarding is complicated by the number and complexity of machines and also by the different uses for individual machines. For these reasons, not all machine builders provide point-of-operation guards on their products. In many cases a point-of-operation guard can only be made and installed by the user after a thorough hazard analysis of the work requirements.

Mechanical Power Transmission Apparatus Guarding

A significant difference between power transmission guards and point-of-operation guards is that the former type needs no opening for feeding stock. The only openings necessary for power transmission guards are those for lubrication, adjustment, repair, and inspection. These openings should be provided with covers that cannot be removed except by using tools for service or adjustment.

To be effective, power transmission guards should cover all moving parts in such a manner that no part of the operator's body can come in contact with them.

Guard Material

Under many circumstances, metal is the best material for guards. Guard framework is usually made from structural shapes, pipe, bar, or rod stock. Filler material generally is expanded or perforated or solid sheet metal or wire mesh. It may be feasible to use plastic or safety glass where visibility is required.

Guards made of wood generally are not recommended because of their flammability and lack of durability and strength. However, in areas where corrosive materials are present, wooden guards may be the better choice.

Chapter 4

Machinery Maintenance and Repair

Good maintenance and repair procedures can contribute significantly to the safety of the maintenance crew as well as to that of machine operators. But the variety and complexity of machines to be serviced, the hazards associated with their power sources, the special dangers that may be present during machine breakdown, and the severe time constraints often placed on maintenance personnel all make safe maintenance and repair work difficult.

Training and aptitude of people assigned to these jobs should make them alert for the intermittent electrical failure, the worn part, the inappropriate noise, the cracks or other signs that warn of impending breakage or that a safeguard has been damaged, altered, or removed. By observing machine operators at their tasks and listening to their comments, maintenance personnel may learn where potential trouble spots are and give them early attention before they develop into sources of accidents and injury. Sometimes all that is needed to keep things running smoothly and safely is machine lubrication or adjustment. Any damage observed or suspected should be reported to the supervisor; if the condition impairs safe operation, the machine should be taken out of service for repair. Safeguards that are missing, altered, or damaged also should be reported so appropriate action can be taken to insure against worker injury.

If possible, machine design should permit routine lubrication and adjustment without removal of safeguards. But when safeguards must be removed, the maintenance and repair crew must never fail to replace them before the job is considered finished.

Is it necessary to oil machine parts while a machine is running? If so, special safeguarding equipment may be needed solely to protect the oiler from exposure to hazardous moving parts. Maintenance personnel must know which machines can be serviced while running and which cannot. "If in doubt, lock it out." Obviously, the danger of accident or injury is reduced by shutting off all sources of energy.

In situations where the maintenance or repair worker would necessarily be exposed to electrical elements or hazardous moving machine parts in the performance of the job, there is no question that power sources must be shut off and locked out before work begins. Warning signs or tags are inadequate insurance against the untimely energizing of mechanical equipment.

Thus, one of the first procedures for the maintenance person is to disconnect and lock out the machine from its power sources, whether the source is electrical, mechanical, pneumatic, hydraulic, or a combination of these. Energy accumulation devices must be "bled down."

Electrical: Unexpected energizing of any electrical equipment that can be started by automatic or manual remote control may cause electric shock or other serious injuries to the machine operator, the maintenance worker, or others operating adjacent machines controlled by the same circuit. For this reason, when maintenance personnel must repair electrically powered equipment, they should open the circuit at the switch box and padlock the switch (lock it out) in the "off" position. This switch should be tagged with a description of the work being done, the name of the maintenance person, and the department involved. A lockout hasp is shown in Figure 68.

Figure 68.
Lockout hasp

Mechanical: Figure 69 shows safety blocks being used as an additional safeguard on a mechanical power press, even though the machine has been locked out electrically. The safety blocks prevent the ram from coming down under its own weight.

Pneumatic and hydraulic: Figure 70 shows a lockout valve. The lever-operated air valve used during repair or shutdown to keep a pneumatic-powered machine or its components from operating can be locked open or shut. Before the valve can be opened, everyone working on the machine must use his or her own key to release the lockout. A sliding-sleeve valve exhausts line pressure at the same time it cuts off the air supply. Valves used to lock out pneumatic or hydraulic-powered machines should be designed to accept locks or lockout adapters and should be capable of "bleeding off" pressure residues that could cause any part of the machine to move.

In shops where several maintenance persons might be working on the same machine, multiple lockout devices accommodating several padlocks are used. The machine can't be reactivated until each person removes his or her lock. As a matter of general policy, lockout control is gained by the simple procedure of issuing personal padlocks to each maintenance or repair person; no one but that person can remove the padlock when work is completed, reopening the power source on the machine just serviced.

Following are the steps of a typical lockout procedure that can be used by maintenance and repair crews:

1. Alert the operator and supervisor.
2. Identify all sources of residual energy.
3. Before starting work, place padlocks on the switch, lever, or valve, locking it in the ''off'' position, installing tags at such locations to indicate maintenance in progress.
4. Ensure that all power sources are off, and ''bleed off'' hydraulic or pneumatic pressure, or ''bleed off'' any electrical current (capacitance), as required, so machine components will not accidentally move.
5. Test operator controls.

Figure 69. Safety blocks installed on power press

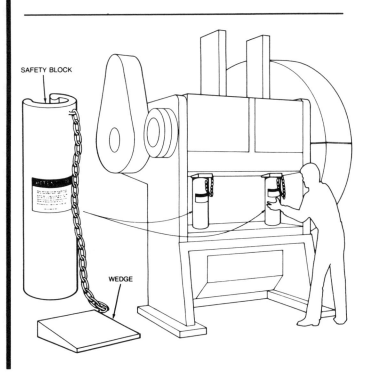

SAFETY BLOCK

WEDGE

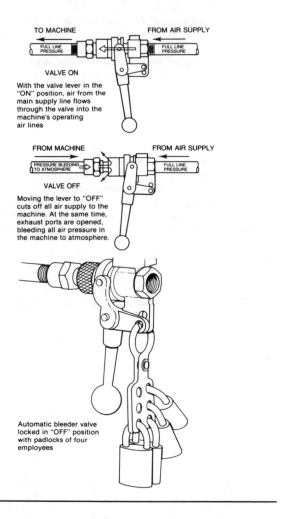

TO MACHINE   FROM AIR SUPPLY

FULL LINE
PRESSURE

FULL LINE
PRESSURE

VALVE ON

With the valve lever in the
"ON" position, air from the
main supply line flows
through the valve into the
machine's operating
air lines

FROM MACHINE   FROM AIR SUPPLY

PRESSURE BLEEDING
TO ATMOSPHERE

FULL LINE
PRESSURE

VALVE OFF

Moving the lever to "OFF"
cuts off all air supply to the
machine. At the same time,
exhaust ports are opened,
bleeding all air pressure in
the machine to atmosphere.

Automatic bleeder valve
locked in "OFF" position
with padlocks of four
employees

Figure 70.
Lockout valve

6. After maintenance is completed, all machine safeguards that were removed should be replaced, secured, and checked to be sure they are functioning properly.

7. Only after ascertaining that the machine is ready to perform safely should padlocks be removed, and the machine cleared for operation.

The maintenance and repair facility in the plant deserves consideration here. Are all the right tools on hand and in good repair? Are lubricating oils and other common supplies readily available and safely stored? Are commonly used machine parts and hardware kept in stock so that the crews aren't encouraged (even obliged) to improvise, at the risk of doing an unsafe repair, or to postpone a repair job? And don't overlook the possibility that maintenance equipment itself may need guarding of some sort. The same precaution applies to tools and machines used in the repair shop. Certainly, the maintenance and repair crew are entitled to the same protection that their service provides to the machine operators in the plant.

Cooperation and Assistance

Safety in the workplace demands cooperation and alertness on everyone's part. Supervisors, operators, and other workers who notice hazards in need of safeguarding, or existing systems that need repair or improvement, should notify the proper authority immediately.

Supervisors have these additional, special responsibilities with regard to safety in the workplace: encouraging safe work habits and correcting unsafe ones; explaining to the worker all the potential hazards associated with the machines and processess in the work area; and being responsive to employer requests for action or information regarding machine hazards. The first-line supervisor plays a pivotal role in communicating the safety needs of the worker to management and the employer's safety rules and policies to the worker.

Sometimes the solution to a machine safeguarding problem may require expertise that is not available in a given establishment. The readers of this manual are encouraged to find out where help is available, and, when necessary, to request it.

The machine's manufacturer is often a good place to start when looking for assistance with a safeguarding problem. Manufacturers can often supply the necessary literature or advice. Insurance carriers, too, will often make their safety specialists available to the establishments whose assets they insure. Union safety specialists can also lend significant assistance.

Some government agencies offer consultation services, providing for on-site evaluation of workplaces and the recommendation of possible hazard controls. OSHA funds one such program, which is offered free of charge to employers in every state. Delivered by state governments or private contractors, the consultation program is completely separate from the OSHA inspection effort—no citations are issued and no penalties are proposed. The trained professional consultants can help employers to recognize hazards in the workplace and can suggest general approaches for solving safety and health problems. In addition, the consultant can identify the sources of other help available, if necessary.

Anyone with questions about Federal standards, about the requirements for machine safeguarding, or about available consultation services should contact OSHA. (See the list of OSHA Regional Offices in the back of this publication.)

Worker Rights and Responsibilities

If you are a worker, you have the right to:

- request an OSHA inspection for workplace hazards, violations of OSHA standards, or violations of the OSHAct (your name will be kept confidential on request);

- have an authorized employee representative accompany the OSHA compliance officer on the workplace inspection;

- confer informally with the OSHA compliance officer (in private, if preferred);

- be notified by your employer of any citations issued for alleged violations of standards at the workplace, and of your employer's requests for variances or for changes in the abatement period;

- contest the abatement time set in any citation issued to your employer by OSHA;

- file a complaint with OSHA if you feel that you have been dismissed, demoted, or otherwise discriminated against for exercising rights under OSHA;

- file a complaint with Federal OSHA authorities if your State agency fails to administer a State program as effectively as required by OSHA;

- ask OSHA about any tests performed in your workplace, the results of inspections, and any decision not to take action on a complaint;

- receive information from your employer about hazards and safety measures applicable to the workplace, OSHA standards relevant to your job, and the record of accidents and illnesses in the workplace;

- ask that NIOSH evaluate and provide information on the substances used in your workplace;

- refuse to work in an imminent danger situation, under certain conditions;

- file suit against the Secretary of Labor if you are injured because of what appears to be OSHA's disregard of an imminent danger situation;

- submit written information or comment to OSHA on the issuance, revocation, or modification of an OSHA standard and to request a public hearing; and

- observe the monitoring and measuring of toxic substances in the workplace if you are exposed, and to have access to any records of your exposure.

You also have the responsibility to:

- read the OSHA poster in the workplace;

- comply with all the OSHA standards, with all requirements of your State-approved plan (if any), and with the employer's safety and health rules;

- report any hazards immediately to your supervisor;

- report to your supervisor any job-related illness or injury; and

- cooperate fully with the OSHA compliance officer who inspects your workplace.

GLOSSARY

Absorption
Toxic substances capable of being absorbed into the human body through the skin.

Acute
A short time period of action measured in seconds, minutes, hours, or days.

ACGIH
American Conference of Governmental Industrial Hygienists.

ASTM
American Society for Testing and Materials.

Caution
A lower degree of hazard than the term Danger or Warning.

CFR
Code of Federal Regulations (OSHA).

Chronic
Long period of action in weeks, months, or years.

Cyanosis
Blue appearance of the skin, indicating a lack of sufficient oxygen in the arterial blood.

Dust
Airborne solid particles with a wide range of size.

EPA
Environmental Protection Agency.

Ergonomics
A multidisciplinary activity dealing with interaction between man, environment, heat, light, sound, tools, and the equipment in the workplace.

Explosion Proof

An electrical apparatus so designed that an explosion of flammable gas inside the enclosure will not ignite flammable gases outside.

Fire Point

The lowest temperature at which a liquid will ignite.

First Aid

The emergency care to a person injured or ill to prevent further injury or death until medical aid can be obtained.

Flammable Liquid

A liquid with a flash point below 100°F.

Flash Point

The lowest temperature at which a flammable vapor-air mixture above the liquid will ignite.

Fume

Very small solid particles formed by condensation of volatilized solid, usually metals.

Gas

Diffuse, form less fluid normally in a gaseous state.

GFCI

Ground fault circuit interrupter commonly used in electrical systems as a protective device.

Hazard

A dangerous condition, potential or inherent, which can interrupt or interfere with the orderly progress of an activity.

Hazardous Substance

A substance possessing a relatively high potential for harmful effects to humans.

Industrial Hygiene

The science devoted to the recognition, evaluation, and control of environmental factors or stresses in the workplace that can cause illness.

Inhalation

Toxic substances capable of being absorbed into the human body through the respiratory tract.

Ingestion

Toxic substances capable of being absorbed into the human body through the digestive tract.

JHA

Job hazard analysis, a study of a given job in order to identify the hazards in a given task.

Mist

Finely divided liquid droplets suspended in air and generated by condensation or by atomizing.

MSDS

Material Safety Data Sheets.

NIOSH

National Institute of Occupational Safety and Health.

OSHA

Occupational Safety and Health Administration.

OSHRC

Occupational Safety and Health Review Commission.

Oxidizer

A material that readily reacts with oxidizable materials (organics — solvents, etc.)

Placard

A sign posted on transport units or in work areas for notification that hazardous materials or substances are present.

Point of Operation

The point of greatest danger to a worker when operating a machine or tool.

PPM

Parts per million.

Risk

The chance of human, equipment, or property loss.

Smoke

Carbon or soot particles resulting from incomplete combustion.

TLV

Threshold Limit Value which is an estimate of the average safe concentration of a toxic substance that can be tolerated over an 8 hour period.

Toxic Substance

A substance that can produce injury or illness through ingestion, inhalation, or by absorption.

TWA

Time weighted average — an average exposure to a hazardous substance over a given period of time which is determined by sampling the worksite.

Vapor

Gaseous form of substances which are normally in the solid or liquid state.

Warning

A term indicating an intermediate degree of hazard in precautionary labeling between danger and caution.

INDEX